SPACE
TOURISM

DO YOU WANT TO GO?

Space Tourism

Do You Want to Go?

by
John Spencer
with
Karen L. Rugg

An Apogee Books Publication

Published by Apogee Books an imprint of Collector's Guide Publishing Inc., Box 62034, Burlington, Ontario, Canada, L7R 4K2, http://www.cgpublishing.com

Printed and bound in Canada

Space Tourism: Do You Want to Go? (First Edition) by John Spencer with Karen Rugg

ISBN 1894959086 – ISSN 1496-6921

CONTENTS

Dedication

To Pamela Douglas

Over the years you have been a wonderful companion, guide, and friend. In your own way, by supporting me, you have made a significant contribution and real difference in the creation of the space tourism movement and industry. Very cool.

Acknowledgments

I very much want to thank Karen L. Rugg for her years of patience, advice, and good writing on this book; and Larry Evans for making it ready for publication while keeping us honest and true to science. I also want to thank Randy Totel for his wonderful computer work on *Destiny*, and Andrea Howe (*www.bluefalconediting.com*) for providing that all-important last set of eyes to make sure we didn't miss anything on the way to Apogee.

Thanks team, I could not have finished this book without you.

And finally, I wish to thank all the pioneers throughout history who did things everyone else said could not be done and who inspired others to follow by creating their own paths. Someday I am confident humanity will design and pioneer its own path to the stars. I am also sure we will go in peace for all humankind.

Karen L. Rugg

Karen is owner of Karlynne Communications. Over the years her clients have included NASA Headquarters, *Space Business International* magazine, the Challenger Center for Space Science Education, the Space Foundation, and ProSpace. She is currently president of Women in Aerospace. Before starting her own company, Karen was director of communications and government relations for the National Space Society. She pushes the boundaries by living in the Washington area, sailing on tall ships, and playing the harp in unexpected places. (*www.karlynne.com*)

Larry Evans

Larry has followed the space program since kindergarten, when his father took him along on business trips to places like Edwards Air Force Base to see the X-15 and other vehicles that would lead the way into space. While serving in the U.S. Air Force, Larry bought his first professional camera system, turning it with great success to fashion and modeling photography. In 1993, he founded Mach 25 Communications to provide space education services to science centers, aerospace companies, and the general public (mach 25 is orbital velocity). Larry is principal writer and photographer for *O.C.Space*, the monthly publication of the Orange County Space Society in southern California, and his photography of Space Shuttle launches and landings has garnered numerous awards.

Preface

Welcome Aboard!

A growing number of open-minded, future-oriented people are joining the space tourism movement. I am one of the first. Our goal is to build a successful space tourism industry over the next several decades, one that enables many people of diverse backgrounds to go off world, to have that very special life-changing experience.

The movement is attracting people from all walks of life, ages, and professions. Most of us are not aerospace engineers, though in recent years some of the best engineers and scientists have joined. Almost everyone involved wants to go someday and feels that by participating we are bringing this quest closer to reality.

However, going is not the only draw. During this long quest, the real personal reward is participating with like-minded people who love to challenge themselves to achieve great goals ... to know you are making a real difference in helping create a more peaceful, healthy, and progressive future for all humankind.

How would space tourism do this? It is the only near-term, off-world endeavor that attracts the interest of millions of non-scientific as well as scientific people, the media, and private investment. A successful private enterprise space tourism industry would finance the development of new technologies and methods of achieving orbit far more economically and safely than would government programs. This success would facilitate large-scale human exploration of our inner solar system, and eventually generate limitless resources of materials, energy, and knowledge that reduce environmental burdens on Earth. Space tourism will lead to the establishment of self-sustaining settlements on the Moon and, eventually, Mars.

Most important, tourists explore. They learn about different people while building understanding and friendships. Orbital tourism is the ultimate boarder-buster experience for those who go. It clearly and dramatically illustrates that every human being is a crew member on a very large and very alone spaceship called Earth. This unique perspective inspires most space travelers with a deep resolve to protect and cherish our home world.

I have been directly involved with real and simulated/entertainment space tourism since the early 1980s. It has been an amazing and rewarding experience to be part of starting a social movement and exciting new industry. I look forward to spending the rest of my career in the service of off-world tourism. I hope you will also join the quest.

I want to go.

John Spencer – Santa Monica, California, May 2004
Space Architect
Founder and President, The Space Tourism Society
www.spacetourismsociety.org

Four Phases of Industry Development

2001	2010	2020	2030	2040

Pioneering Phase ↑

Wealthy private citizens and lottery winners

ISS conversion into first private orbital yacht

Exclusive Phase

Second generation of orbital yacht

Begin orbital sports activities

The Space Guard Service begins operations on orbit

Mature Phase

Third generation of orbital yachts starting with *Destiny*

Development of first orbital yacht club — *Eagle*

Orbital Yacht Racing

Mass Market Phase

Space cruise ships and lunar cruising

First sports stations and orbital ports

The evolution of the Earth-based immersive space/lunar/Mars simulation industry

FOREWORD

Do you want to go? Well of course, how could anyone not want to go? Such was my reaction when I first read the subtitle to this book. Of course I had been fortunate enough to have flown to space not once, but three times. Accordingly, I knew of marvels beyond description. Certainly my trips weren't vacations, but even during very demanding and pressure-packed workdays, astronauts still enjoy the singular and fabulous sensations of weightlessness coupled with unbelievable views. I knew firsthand the liberating sensation of floating freely lighter, literally, than a feather and had experienced being an out-of-this-world gymnast, high jumper, pole-vaulter, and Pele-style over-the-head-kicking soccer star in my weightless unfettered imitations. I'd been enthralled watching huge, perfectly spherical bubbles of water held together by only surface tension, had chased Goldfish crackers around the cabin while humming the *Jaws* theme song, and had played catch with crewmates using another curled-up crewmate as a ball. Who wouldn't want to experience such fun?

Even better than the fun of weightlessness, I had reveled in glorious and awe-inspiring, ever-changing views of our home planet from orbit. Seen how India is crashing into Asia to thrust up the Himalayas, the roof of the world, and spied the elusive Mt. Everest summit from 170 miles right overhead. Witnessed that in the Caribbean there are more shades of blue, jade, and turquoise than I could have ever imagined. Pondered the richness of ancient history as I drifted silently over the Mediterranean. Been especially touched by the obvious fertility, unparalleled variety, and patriotic personal connection when flying over America, a land that one view from orbit testifies is truly a promised one. I had been washed in the light of those breathtaking orbital sunrises and marveled at how thin and fragile our atmosphere looks on edge off on the curved horizon. Seen lightning in the tropics popping off like flashes in a dark auditorium, with the occasional shooting star plunging to its fiery orange demise below me. Who would not want to see such wonders of wonders?

In discussing my experiences with thousands of people through the years, although I can't understand how anyone would not want to go to space, I have observed that there is very little middle ground. Either people are like me, "in a heartbeat, I've wanted to do it since I was five, it's got to be the greatest adventure ever," or they are just the opposite. "You'd never catch me climbing on to one of those rockets," or "I'm claustrophobic, I couldn't stand being cooped up like that," or any one of hundreds of similar expressions that trade the deep intangible benefits for avoiding risk, danger, or discomfort. If you are reading this book, chances are you are in the former category. And, as John explains with such fervor and vision, in our future the opportunities to really go to space will expand. Much as air travel grew from a dangerous, irregular proposition to virtual bus lines in the sky, access to space will eventually become safer and more common. You may indeed have the chance!

Twenty-five years ago, using the words "space" and "tourism" together elicited virtually only laughter except from long-range, optimistic, visionary thinkers like John. Now we have had real space tourists, albeit to date just a few very wealthy individuals, but the door is open, and the entire array of space tourism lies before us. From one end of children enjoying the suspense of a countdown and the thrill of launching model rockets, to immersive simulations and parabolic flights giving repeated 25-second periods of weightlessness, to soon-to-be flying suborbital reusable launch vehicles and orbital experiences, it's a fascinating, mind-expanding spectrum of adventures that awaits you! Here you have a guide book.

While the exact timetables and flow of the future are never exactly predictable, this book has indeed, I believe, captured the essence of where we are headed and what may happen. Enjoy the read, ponder the possibilities, catch the vision. Let this work grab your imagination and open up new vistas to you. As you do so, keep one thing in mind: we humans almost always tend to overestimate progress in the short term and underestimate progress in the long term. With that in mind, who knows where we'll be, what we'll be doing, or what future awaits in space even by the end of our lifetimes, let alone 100 or more years from now. So strap in, hang on, and enjoy the ride!

Rick Searfoss, Colonel, USAF Retired
Commander: STS-90, Pilot: STS-58 & 76

Former astronaut Rick Searfoss, Colonel, USAF Retired, commanded the STS-90 *Neurolab* Space Shuttle mission on *Columbia*. He also piloted two other spaceflights, including a mission on *Atlantis* to the Russian *Mir* space station. Professionally he now divides his time between corporate speaking on leadership, teamwork, and innovation, and consulting for numerous private space ventures. He can be found on the web at *www.astronautspeaker.com*. With his fervent belief that the world of entrepreneurial space activities is where the real pioneers are today, Rick hopes to yet again fly to space as a pilot of future suborbital reusable launch vehicles to take real space tourists on the ride of a lifetime.

PROLOGUE
THE VISION

"I Love Space."
Dennis Tito, April 30, 2001
First private space traveler upon entering the
International Space Station, in Earth orbit.

ust 100 miles above where you are right now, reading this book, lies the start of a fantastic voyage. A voyage you can personally experience someday. A voyage into space. Actually, he edge of space in low Earth orbit (LEO) is less than 100 miles from anywhere on Earth. Closer to you perhaps than a major Earth-bound city, shopping mall, or theme park.

Since the early 1980s I have introduced the concept of space tourism to a wide variety of people from around the world. I have found that more than 90 percent of those people believe space tourism will become a new industry. They just want to know when can hey go and for how much.

Today, the pieces are slipping into place that will someday make it possible for large numbers of private space travelers — tourists — to visit LEO and beyond.

The next generation of privately developed and operated reusable orbital access vehicles will dramatically bring down the cost of going off world while significantly increasing safety. Lotteries, contests, and sponsorships will widen the opportunity for the average healthy person to go.

The space tourism industry officially began on Saturday, April 28, 2001, at 12:37 a.m. PST when Dennis Tito, a wealthy Los Angeles businessman and former aerospace engineer, and two cosmonauts, Commander Talgat Musabayev and flight engineer Yuri Baturin, lifted off from the Baikonur Cosmodrome in Kazakhstan on an eight-day mission o the the International Space Station. On his return to Los Angeles on May 9, Tito encountered a media frenzy and a mayor who proclaimed "Dennis Tito Day."

More than 10 years earlier, on December 2, 1990, Toyohiro Akiyama, a Japanese journalist for the Tokyo Broadcasting System, lifted off from Russia aboard a *Soyuz* rocket on an eight-day day trip to the *Mir* space station. From *Mir* he broadcast his experiences and became a national hero. The Russians were paid $12 million for his ticket. In 1991 the Russian space program sold a second trip to *Mir*, valued at $10 million to a British company. And 13,000 Britons responded to advertisements in the press to send Britain's first person o space. The winner was Helen Sherman, a 27-year-old chemist. She blasted off from Russia on May 18, 1991, spending eight days off world.

These space tourism pioneers will always be remembered as the first to go. However, Dennis Tito will retain the title of "first space tourist" because he was the first private citizen o pay for his dream trip with his own funds. Tito was free to set his own agenda of taking pictures, listening to music, and just floating. He was there for the experience — nothing more, nothing less. In April 2002, Mark Shuttleworth became the second wealthy individual to pay for a trip to Earth orbit. In 2003, two more space tourists may have flown had we not lost the Space Shuttle *Columbia.*

These flights are just the beginning. According to Dennis Tito, among the 80,000 Americans whose personal net worth is at least $10 million, there are dozens who are interested and capable of paying for trips to space. Those dozens grow to hundreds if the payments can be spread over time. In addition, there are hundreds of huge corporations who may be willing to spend millions to send their people to space as rewards and for promotional purposes.

We should never underestimate what these space tourism pioneers achieved. They proved that a space traveler does not need to be rubber-stamped by a government agency to get to space. Their trips are the first steps. Now we can fast-forward to what the space tourism experience can become.

You Will Never Forget

Experiencing your first countdown and liftoff. Seeing Earth for the first time from LEO. Floating free in zero gravity.

Space cruise lines will offer space tourists a wide variety of orbital and lunar cruises. Relax on a one-week orbital cruise aboard an elegant space cruise ship christened *Lady of the Stars*, *Orbital Ecstasy*, or *Star Princess*. These majestic spaceships will offer the highest quality human, artificial intelligence (AI), and robotic crew service. Spend quiet hours enjoying the beautiful views of Earth as she silently glides past at 17,500 miles per hour, giving you a crystal-clear sunrise or sunset every 45 minutes. Dance in zero gravity. Swim in spheres of warm, colored water. Enjoy games, sauna, exercise, massages, and world-class entertainment. Shop for rare products made off world. Experiment with zero gravity hair styles and wear the latest space fashions, body paint, and accessories. Take private or group space walks ("floats"). Enjoy the finest dining and, for the romantic at heart, try zero gravity lovemaking in the privacy of your orbital stateroom.

From the orbital space yacht club, *Eagle*, cheer for your favorite space super yacht — *Challenger*, *Intrepid*, or *Quasar* — during a half-million mile lunar regatta swinging around the Moon and back to Earth. The yacht's colorful solar sails capture light particles in the solar wind as they race to win the coveted Solar System Cup. Post-race, celebrate with the winners among some of the most exciting and affluent people to be found off world.

If your destination is the Moon, you will never forget seeing for the first time the Earth disappearing beyond the rim of the Moon, then reappearing in your first Earthrise; seeing your boot prints on the Moon's gray, dusty surface and knowing they will still be there in a million years; picking up a Moon rock and tucking it into your spacesuit pocket for the trip back to Earth; building a lunar sandcastle with soaring arches and tall thin towers in one-sixth gravity inside a golden dome filled with trees and flowers; flying like a bird with silver wings strapped to your arms. The grand finale? A blazing reentry through Earth's atmosphere to a gentle landing.

Though you may eventually visit space and the Moon many times, the memories of hat first voyage away from Earth will always be treasured. Going off world will be a life-changing experience. A deeply personal experience that forces you to stop, to think about ife, love, and the future.

The government-employed astronauts and cosmonauts who have traveled to space ave certainly not experienced the luxurious, free-form experiences I have just described. Those experiences lie decades in the future. However, many of them have returned with a greater appreciation of the beauty and fragility of our home world — our one and only spaceship Earth. Some have said it is almost frightening to see just how thin the blue line of atmosphere that marks the horizon. Many say they have taken an extraordinary journey not just into outer space, but into private inner space, where they have gained important new insights. Many seek to return to reexperience that heightened sense of awareness and connection to the cosmos.

An Industry's First Steps

Today's Earth-based space experiences can be best described as Cro-Magnon. They are the early ancestors of the space experience described in the preceding section, a taste of what is to come.

Almost any one of us today can taste space tourism in Earth-based experiences:

- Fly in a special aircraft that produces multiple periods of 30 seconds of zero gravity as it flies through a rollercoaster flight pattern (parabolas).
- Attend space camps and participate in simulated Space Shuttle and Space Station missions.
- Visit space museums, space theme parks, and the Kennedy Space Center to witness the power and majesty of a space launch.
- Join the growing number of space entrepreneurs who are building space-based businesses and tourism companies.
- Join the growing number of space tourism societies, institutes, and other nonprofit groups researching and designing the space tourism industry.
- Take a growing number of classes and attend space tourism expos, conferences, and workshops.
- Read more books on off-world tourism or write one. Write or produce a television series or a movie on space tourism.

Present-day space tourism companies are laying the groundwork for making the space tourism industry a reality through simulation centers. Other steps that will be occurring in the near future include:

By 2008, companies could be offering suborbital adventure flights in a new generation of private vehicles. These exciting trips would offer 10 minutes of zero gravity.

By 2010, the International Space Station may be completed. Companies are now striving to sell multimillion-dollar tickets with several credible candidates lining up to fly once the shuttles begin flying again. Nationwide lotteries are also in the planning stages as well as promotional programs to send more private citizens.

By 2012, private suborbital vehicles could fly from California to England in just 45 minutes. They would offer 20 minutes of zero gravity.

By 2015 the Space Station could be purchased by private enterprise and converted into a first-generation orbital yacht.

By 2025, the second generation of orbital yachts could be operational, offering first class, off-world services for six to eight voyagers plus crew.

By 2030, the third generation of orbital yachts — the first orbital super yachts — take 20 voyagers plus crew on one- to two-week orbital voyages in fantastic luxury. Films and television or "vid" shows could be filmed off world in mini movie studio stations. Orbital tourists will cheer as the first human missions ships leave for Mars.

By 2040, orbital cruise ships large enough for 100 passengers, offering all the amenities of luxury ocean liners, could be operational. A great competition between space cruise lines could inspire extraordinary technical advancements as they strive to build the largest and grandest ships. Orbital yacht races like the America's Cup on Earth could stimulate huge sponsorships.

By 2050, more than 100,000 people per year (270 per day) could be going off world for vacations and to attend sporting events. Adventure vacationers will head to the Moon to stay at a lunar resort/spa and sports stadium. Extreme sports corporations will sponsor crewed rover races around the Moon's equator, stimulating huge sponsorships, awards, and technological and operational advancements.

And this takes us just to the year 2050. Where will off-world tourism be in the year 2100? It is anyone's guess, but I am highly confident the space tourism experience 100 years from now will be far more advanced and amazing than even I can imagine.

Isn't This Just Pie in the Sky?

Yes! A wide variety of pies, caviar, and the finest wines from Earth's vineyards will be served aboard fleets of privately owned and operated ships and facilities, both in orbit and on the Moon.

Seriously, space tourism is inevitable because it offers an unparalleled challenge, accompanied by the promise of great prestige and profit, to be won by those who take it on. Political leaders, think tank gurus, scientists, advocates, and entrepreneurs agree that humans traveling to space in larger numbers is not a matter of "if," but only of "when."

Just 100 years ago, only rich male adventurers traveled on African safaris, facing real danger and isolation. Many of these explorers became world famous and many died trying. Today, tourists take their entire families on African photo safaris, staying at five-star resorts on wildlife preserves. Since 1997, one million people every year take submarine passenger rides. Thousands visit the Arctic and Antarctic on expedition cruises. Several people every year pay more than $30,000 each to take a deep-dive submersible five miles beneath the ocean to pay their respects to the *HMS Titanic*. There are so many people climbing Mt. Everest that there are traffic jams. In 50 years 1,300 have made the summit and 200 have died trying.

This passion to explore new horizons and to experience unfamiliar yet unforgettable tastes, sights, and sounds are within each of us. Millions of people around the world dream of going off world. Several market surveys in the United States and abroad clearly demonstrate that a large international interest and audience exist. In the United States alone, 42 percent (almost 80 million) of respondents to a survey said they would like to take a space vacation someday.

An orbital gold rush has begun. Private enterprise is clearly at the head of this new industry. Great dramas will unfold and huge fortunes will be made as we pioneer this new design, development, and experience frontier. The future equivalents of Walt Disney, Howard Hughes, and Bill Gates are working today to create the space tourism industry of tomorrow. You could be one of them!

My confidence grows every year that the space tourism movement and industry will prosper over the next decades as more billionaires and wealthy individuals join the quest. And they are joining. In the June 2003 issue of *Wired* magazine, writer Carl Hoffman penned an article titled "The Countdown Begins: The Next Space Race Will Be Won by Rich Geeks with the Right Stuff." Hoffman profiled the many wealthy individuals who have started companies to develop the next generation of private rockets and space planes. In a sidebar titled "Amazon Enters the Space Race," writer Brad Stone noted that Amazon founder Jeff Bezos, whose net worth is estimated at almost $3 billion, stated in a 1982 *Miami Herald* article that "he hoped to one day put space hotels, amusement parks, and yachts in orbit." In fact, for the past few years, Bezos has been funding a very quiet research and development company based in Seattle called Blue Origin (*www.blueorigin.com*). One of their projects is a spaceship tentatively called *New Shepard*, which would carry seven space tourists into orbit several times each week.

On December 17, 2003 — the 100th anniversary of powered flight — Scaled Composites, Inc., founded by Burt Rutan, achieved a milestone in the quest for private spaceflight. Their aircraft, *White Knight*, dropped a second aircraft called *SpaceShipOne*, which then fired a rocket to become the first private aircraft to break the sound barrier. Their goal is to win the Ansari X Prize Competition and its $10 million prize by being the first private company to fly three people on a suborbital flight in a reusable spaceship. That day it also became official that the cofounder of Microsoft, Paul Allen, one of the richest men in the world, was funding Rutan's effort with $25 million of his personal fortune.

Why Is Space Tourism So Important?

A successful private space tourism industry will generate the financing and attract the talent required to design and develop the infrastructure, new technologies, and fleets of orbital access vehicles that will economically support all endeavors vital to the long-term prosperity of the human race. These endeavors include human exploration of the Moon and Mars, expanded scientific research, industrial development, and military security.

The new national vision for U.S. space exploration released by President Bush in January 2004 acknowledged that "Direct human experience in space has fundamentally altered our perspective of humanity and our place in the universe." This vision continues to state that one of its chief goals is to "extend human presence across the solar system, starting with a human return to the Moon by the year 2020, in preparation for human exploration of Mars and other destinations."

Space tourism, spurred by NASA-led human exploration missions, will facilitate the construction of human colonies in space and on other planets. Or, as I believe, if some disaster threatens Earth's population, 50 years from now space tourism will help to weave a safety net for our species.

First Things First

The job ahead of us is to simply take our cue. To, as did the engineers in that exhilarating scene from the film *Apollo 13*, toss every idea we have onto the table and, as a team, build solutions for the challenge. To put the best minds and talents to work creating a space experience that is, for now, mostly here on Earth. But it will not be for long. Soon we will be able to replicate the simulated trips, journeys, and experiences in real space. Getting there will call for all talents — teaching, design, engineering, physics, music, interior decorating, medicine, cuisine, athletics, art, and many others.

This book provides a model for moving forward — a model many of us use every day to advance the value and relevance of space tourism. Most important, this book is a primer for deciphering your role in this grand quest. You *do not* have to be a rocket scientist to join. You *can* make meaningful contributions through your ideas, skills, and energy. You *can* make a difference in expanding the space renaissance and, at the same time, have a wonderful life working with inspired fellow space pioneers.

Do you want to go?

INTRODUCTION
THE SPACE RENAISSANCE

THE SPACE RENAISSANCE

"The Lights Are On"
One of the Space Shuttle Endeavour *astronauts in December 1998 as the first two modules of the International Space Station connect in Earth orbit and the power is switched on.*

Introduction

Renaissance: A revival or rebirth. A period of expanded intellectual or artistic achievement or enthusiasm.

That is what is happening now in the hearts and minds of anyone who has anything to do with private space development. This revival — complete with raised voices, hands clapping, and ringing choruses of "Alleluia!" — has to do with space tourism, the most important development in the future of space exploration. Against a backdrop of diminished expectations for the International Space Station, underutilized satellite constellations, hexed boosters, and diminished launch markets, shines the brightest spot in the newly minted history of public space travel — the April 28, 2001, flight of private citizen Dennis Tito.

Yes, we are right at the beginning of a space renaissance. And it is happening because we are finally getting it — that this revival is a private show. It is fueled by you and me, not by mega-government agencies, not by skunk-black secret projects.

Finally and at last, private companies — private citizens — are participating and driving the planning, design, development, financing, and operation of space simulations, orbital facilities, and spaceships. Finally and at last, we are wresting the development of space tourism from the hands of those who are not going to do a thing about it and placing it in the hands of those who can. You can now play a role in creating new concepts and new companies and, eventually, leading the further expansion of humanity into space.

Finally and at last, we get it. It is up to you and me to make available more flights like Dennis Tito's. To put a face-splitting grin on many more faces and hear "I love space!" from many more lips. That's what this book is about — following a unique vision to create a unique industry and having a damn good time doing it!

The Pioneering Personality

To better understand your role in the creation of the space tourism industry, it is important to understand the types of people with whom you will be working. The first steps to this space renaissance are not being taken by governments or large aerospace corporations. They are being taken by inspired and knowledgeable private citizens with new ideas for space businesses with the strength and dedication to develop them. While there are many aerospace engineers, retired astronauts, and real rocket scientists participating in creating the space renaissance, the majority of these new pioneers do not have aerospace or engineering backgrounds. We come from a wide array of disciplines, including architecture, banking, filmmaking, marketing, promotion, entertainment, sales, psychology, medicine, travel, law, multimedia, writing, art, and education.

Our common bond is a strong belief that space development is important to the healthy future of humanity. We are also excited by extreme challenges, the prospects of wealth and fame, and the joy of participating in something with real meaning and purpose. (Sound like you?)

Many have compared this resurgence in pioneering instinct and the desire to populate a new frontier to the evolution of the aviation industry in the late 1940s through the 1960s. During that brief 20-year period, private commercial airlines experienced stunning technological advancement, huge international growth, healthy competition, and diversity. Another analogy is the Great California Gold Rush in the mid-1800s. As soon as the railroad stretched across the entire continental United States, and gold was discovered in California, tens of thousands of people made the trek west seeking their fortunes. Those analogies have some merit, however, they do not present the most appropriate model for space tourism development.

For the emerging space tourism industry, the best analogy is the evolution and impact of the cruise line industry. Beginning in the 1600s with wooden sailing ships carrying a few passengers and mail from Europe to the New World, the shipping lines grew into large companies pioneering new technology and competing to build the world's greatest ocean liners carrying thousands of passengers.

The cruise lines created the means for the greatest migration in human history, enabling millions of immigrants to cross the Atlantic and start new lives in America. The space cruise lines of the future may well be the means through which millions of people emigrate from Earth to the Moon, to huge colonies built in space and even on Mars.

Over time, the cruise lines became much more than a means of mass transportation. With years of operation and service behind them, cruise lines began to focus on creating experiences to maximize the unique nature of their travel. Do airlines do this today? Certainly not for the masses. Do railroads? No. Will space cruise lines? Yes. And therein lies the difference. As we saw with Dennis Tito, it was the experience that made his training and sacrifices worthwhile. He wanted the experience of being in space. Remember, "it's the experience, stupid."

Today we are striving to put our own version of a cruise line into orbit with a growing diversity of new designs for lower-cost reusable launch vehicles, aerospaceplanes, and other advanced concepts. A multitude of ideas for orbital and lunar industries inspire visions of an orbital gold rush using a transportation system as reliable and trailblazing as the railroad. The new gold in the sky consists of communications satellites, scientific research, enhanced security, and by far the largest potential market of all: space tourism.

Renaissance Roots Are Recent

On October 29, 2000, there were no humans in Earth orbit. None aboard the Russian *Mir* Space Station. None aboard the International Space Station. None in a Space Shuttle or a Russian *Soyuz* capsule going to or returning from orbit. On that day humanity was a planet-bound species.

Thankfully, the zero population count in space was short-lived. On October 30, 2000, at 11:53 p.m. PST, a Russian rocket lifted off its launch pad in Kazakhstan carrying three space pioneers who would become the first crew of the International Space Station (ISS). Their mission was called *Expedition 1*. The commander of the mission was U.S. astronaut William M. Shepherd, a former U.S. Navy Seal. His crew members were Russian cosmonauts Yuri Gidzenko and Sergei Krikalev.

After two days, at 1:21 a.m. PST, approximately 250 miles above Earth, the *Soyuz* capsule docked with ISS. One hour later, *Expedition 1* entered the station for a four-month shakedown cruise, including coordinating two Space Shuttle missions tasked with adding power systems and the *Destiny* laboratory module to the station.

But it was the name-calling that made the most news during this mission — name-calling in a positive light, that is. When Commander Shepherd radioed to the Russian mission command center and gave the call sign, *Alpha*, the NASA Administrator at the time, Dan Goldin, sat stunned while Russian associates around him broke into loud applause.

What we had just witnessed was a bit of poetic justice. Back in the early 1990s, Shepherd became NASA's program manager for the post-Reagan space station design — the one we have today. That design was called, coincidentally enough, *Alpha*. Years later, as Shepherd began training for this mission, he asked Goldin for permission to name the station *Alpha*. Permission was denied. But once *Expedition 1* was on board the station, Shepherd tried again. In the first communication to Russian Mission Control, he asked permission to use *Alpha* as a call sign. This time Goldin said yes, but only for Shepherd's crew. Too late! The horse was out of the barn. Today, the station is unofficially known as Space Station *Alpha*.

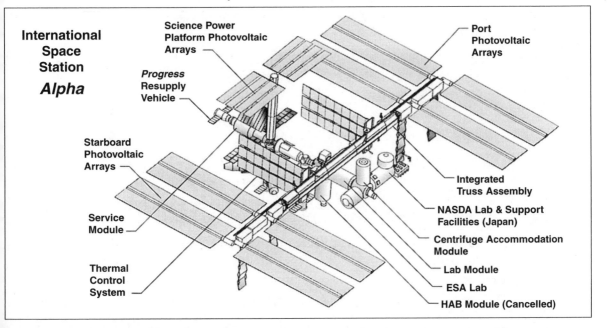

International Space Station **Alpha**

Science Power Platform Photovoltaic Arrays

Progress Resupply Vehicle

Port Photovoltaic Arrays

Starboard Photovoltaic Arrays

Service Module

Thermal Control System

Integrated Truss Assembly

NASDA Lab & Support Facilities (Japan)

Centrifuge Accommodation Module

Lab Module

ESA Lab

HAB Module (Cancelled)

Officially, the International Space Station is the most complicated and expensive international project in human history. In development since the early 1980s, it will cost more than $100 billion to complete. A 16-nation consortium, led by the United States with the Russians as a major partner, is building the station on orbit, 240 miles above the Earth. Originally designed to house seven international crew members for three-month tours of duty, budget overruns have limited the crew complement to three (two during the *Columbia* stand down).

The International Space Station is a complex collection of pressurized modules, each about 15 feet in diameter and 30 feet long, which provide living, research, and operations areas for the crew. The modules are connected to each other and to a large truss structure that also anchors all the solar panels that store the energy required to power the station. Rescue vehicles, docking ports, and other devices form this 100-yard-long facility. Approximately 50 Space Shuttle and Russian *Soyuz* and *Progress* flights will be required to complete construction of ISS by 2010. It is now one of the brightest objects in the night sky, cruising from horizon to horizon during its perpetual 90-minute orbits.

Why all this history about ISS? Because those of us in the private space enterprise community and the space tourism movement still consider the station a renaissance rabble-rouser, a starting point in orbit, a true "place in space." Think of the completed International Space Station as kin to the wooden Army forts of the old Wild West. Built by the government and staffed mostly by government employees, the forts eventually became hubs of private transportation, commerce, and communication. Small towns grew up next to the forts to provide their garrisons with services and to, in turn, be protected by those same garrisons. Those small towns grew into small cities, then larger cities, and finally, the forts were no longer needed.

Think of ISS as a space fort on the edge of the space frontier. Today it is a hub for *Soyuz* and Space Shuttle missions, as well as for scientific research. Already, several private companies are providing critical services to *Alpha*. With an operating life of 15 to 20 years, the station has some commercial interests exploring the possibility of buying it after decommissioning, similar to the way decommissioned naval vessels and military bases have been sold to private companies.

One of the most exciting proposed uses of a retired and then revitalized and expanded private space station would be that of a first generation orbital space yacht. A "place in space" used by the rich and famous and those lucky few who win lottery tickets for the vacation of a lifetime.

Renaissance Reveals Strange Bedfellows

For all of NASA's vaunted vision and *Apollo* legacy, perhaps the most surprising revelation of the recent space renaissance is the degree to which the Russian space program has been much more open to commercial space ventures than NASA. Just look at the media reports surrounding Dennis Tito's April 2001 flight. Who was cheering him on, and who was throwing up barriers?

One week before Tito flew, I debated retired U.S. astronaut Dr. Jerry Linenger on

The Today Show. I took the position that Dennis should be allowed to fly and stay on the station. Jerry, who had spent five months on *Mir*, took the NASA position that Tito should not fly until the station was completed. NASA's negative position was a huge public relations disaster and angered their most important partner, the Russians. Jerry was in a difficult position because he supports space tourism, but had to toe the NASA party line. At the end of the debate he said, "I wish everyone could go up there. I want you to go up there. You would love it. It's the grand view of the world. It's absolutely incredible. I think it would change human beings and the way we look at each other, and realize we are all in this together, but I can't snap my fingers and get you up there."

The Russians have become the "cowboys" of Earth orbit. As many times as *Mir* (Russian for "peace") tried to buck them off with fires, bad air, decaying orbits, runaway supply ships, and faulty computers, those crafty astronauts and cosmonauts cobbled things together time after time. The Russians are clearly embracing the capitalist mantra as they seek to make money from their orbital ventures. They have the only space program to date that has made millions of dollars by selling seats on their rockets to private companies and private citizens. They even filmed a Pepsi television commercial and birthday greetings inside their *Mir* Space Station and Radio Shack commercials inside the Russian part of ISS.

Since the late 1990s the Russians have leased parts of their cosmonaut training facilities near Moscow to tour groups, and were the first to offer "zero gravity" rides in their cosmonaut training aircraft for about $5,000 per person. This unique service was developed and managed by Space Adventures, Inc., the world's first space tour operator.

In early 2000 the Russian space agency that òperated *Mir* formed a commercial company with American investors called MirCorp, and took operational control of *Mir* for commercial uses including tourism. Their business plan was designed to attract major international corporate investment and media sponsorships, and to sell training and passage for eight- to ten-day stays on *Mir* to whoever could afford the multimillion-dollar tickets and the months of training in Russia. Their long-term goal was to convert *Mir* into the first private enterprise port in orbit.

On April 6, 2000, two cosmonauts docked their *Soyuz* capsule with *Mir* to boost the station into a higher orbit and perform routine maintenance and repairs. This multimillion-dollar mission was the first crewed mission fully financed by private enterprise. However, due to political, economic, and technical problems, *Mir* was deorbited in a spectacular light show in the South Pacific near Fiji on March 23, 2001.

The development of orbital enterprise is no longer limited to the United States and Russia. China joined the exclusive club of human spaceflight in 2003, becoming the third nation to send its citizens into orbit aboard its own rockets. They call their space travelers "taikonauts," the Chinese word for space. France, Japan, and India have expanding unmanned space development programs. They have successfully launched their own military and commercial satellites, and have had their astronauts fly aboard the Space Shuttle. With their family of Ariane rockets, France is successfully competing with the United States in the satellite launch industry. Canada, England, Australia, Israel, Brazil, and others are developing their own rockets to launch satellites. Reputable developers in England and Japan are researching and designing tourist passenger vehicles. They could become real competitors as the space tourism industry grows.

Space Entrepreneurs Give Birth to Industry

Since 1996, more than $1 billion has been raised in venture capital and equity participation by more than a dozen small, new, privately owned space vehicle development companies. They are all dedicated to designing and flying their own lower-cost reusable rockets and aerospaceplanes. The driving force behind private development of a new type of reusable vehicle is that their reusability and shorter turnaround schedules will lead to lower cost access to space. The Space Shuttle is a reusable vehicle, yes, but its months-long turnaround time and high maintenance requirements boost the cost of a single mission into the hundreds of millions of dollars.

Private pioneering companies include: Universal Space Lines with its *Space Clipper* vertical-takeoff-and-landing rocket, Kelly Space and Technology with its *Astroliner,* Bristol Spaceplane with their *Ascender* spaceplane, Vela Technology Development with their *Space Cruiser*, Scaled Composites with their *White Knight/SpaceShipOne*, Kistler Aerospace with their *K*-series rockets, XCOR Aerospace, SpaceX, Pioneer Rocketplane, and several others.

The majority of these vehicles are currently not designed to carry human passengers. In most business plans, the market for space tourism is not even addressed. However, behind closed doors or in space conference hallways, the discussion almost always leads to space tourism.

The biggest challenge these real rocket scientists and space entrepreneurs face is not a lack of experience, innovative ideas, passion, market need, or proven technology. It is to develop credible business and operational plans that convince the investment community that the orbital gold rush is real and coming soon enough on the return-on-investment (ROI) timeline to be of value.

Most of the new rocket companies will fail or will be purchased by larger companies. That is the natural process of building new industries. At one time more than 100 railroads competed for passengers and freight. More than 2,000 car manufacturers produced product, as well as many dozens of airlines. Survival of the fittest and resulting consolidation are the natural order of things.

What is most remarkable about private citizens who start their own space access and service companies is that they are challenging the old perception that only countries or huge aerospace corporations can develop access to orbit. Private enterprise democratizes the process of traveling off world.

An Ill-Fated Start for Citizen Access

The programs and technologies developed by the National Aeronautics and Space Administration (NASA) have comprised the United States' only space program to date. While the *Apollo* era spelled success for the agency and for America, its astronauts and technologies were largely out of reach to the public.

That changed when a new chapter in spaceflight began on April 12, 1981, with the Space Shuttle *Columbia*'s two-day maiden test flight — two years behind schedule but, ironically, on the same date 20 years earlier when the Soviet Union launched the first human, Yuri Gagarin, into Earth orbit.

The *Columbia* was flown by test pilots John Young and Robert Crippen. Both had been frustrated by the long delay. Young was heard to comment that he hoped the shuttle flew before Young and Crippen were "old and crippled." The flight was a huge success.

With confidence growing in the Space Shuttle fleet, more crew members were added until a maximum of eight flew at one time (STS-61A, October 30, 1985), making that year an exciting one for proponents of space tourism. NASA's administrator had begun research into how the agency could fairly select private citizens for flights on the Space Shuttle. NASA wanted to show the world how safe and easy spaceflight had become in their wonder vehicle. This program for sending a teacher, journalist, or artist into orbit to better communicate the experience of being in space was known as the Spaceflight Participation Program. The first participant was to have been a journalist, however, the Reagan Administration asked that the participant be a teacher instead. During the application period, NASA received more than 11,000 applications. From those, the agency selected two individuals per state, then narrowed those selections down to one winner — Sharon Christa McAuliffe.

The world soon learned how passionate this wonderful and enthusiastic woman was about space exploration and teaching. She proved she had the "right stuff" during training and simulations. She planned to teach classes from orbit. She personified the space tourist — someone who would travel to space for the experience, not for the science.

Before Christa's turn, half a dozen other non-astronauts flew as payload specialists on four Space Shuttle missions as guests of the United States — a U.S. congressman, a U.S. senator, a Saudi prince, and a McDonnell Douglas engineer. There are about 120 astronauts in the current NASA astronaut corps. Within this corps of professional astronauts are three crew designations: commander, pilot, and mission specialist. A fourth term, payload specialist, defines an astronaut who flies for NASA but who does not require the full two years of professional training. Even here Christa's case was unique — she would not be a payload specialist. She would be purely payload — the world's first private citizen in space.

On January 28, 1986, the 25th shuttle mission began with exuberance and ended in despair when the Space Shuttle *Challenger* exploded 73 seconds after liftoff, killing its seven-member crew. On that day America's space machine ground to a halt. Investigations ensued and the Spaceflight Participation Program was canceled. It was nearly three more years before the launch of the Space Shuttle *Discovery* in October 1998 lifted our hopes back into orbit, but the crew were all seasoned veterans.

The *Challenger* tragedy was to the space tourism movement a far more devastating tragedy than the *Titanic*'s sinking was to the cruise line industry. Both tragedies led to significant improvements in, and attention paid to, safety. But it was a long ten years before NASA again flew a semi-private citizen, senator and former astronaut John Glenn, into orbit as a payload specialist, and only then after powerful political pressure was applied. On October 29, 1998, the world, including President Clinton and the First Lady, 2,000 journalists from around the world, and almost one million people in the Kennedy Space Center viewing area watched as STS-95 left Pad 39B. Hundreds of millions of people worldwide watched this historic and exciting event on television and on the Internet.

John Glenn was 77 years old at the time of his mission, justifying his presence as a benchmark for stockpiling medical data revealing the effects of weightlessness on older humans. On their return, Glenn and his crew received the largest ticker tape parade held in New York City since the return home of troops from *Desert Storm*.

On February 1, 2003, nearly 22 years after her first launch, the *Columbia* and her brave crew of seven were lost only 16 minutes from landing on Earth after a 16-day science mission. It was the 113th flight of the Space Shuttle fleet. On February 3, 2003, Evelyn Husband, wife of *Columbia* Commander Rick Husband, read the following statement to the media on behalf of all of the families of the ill-fated crew:

"On January 16th we saw our loved ones launch into a brilliant, cloud-free sky. Their hearts were full of enthusiasm, pride in country, faith in their God, and a willingness to accept risk in the pursuit of knowledge — knowledge that might improve the quality of life for all mankind. … *Columbia*'s 16-day mission of scientific discovery was a great success, cut short by mere minutes. Yet it will live on forever in our memories. We want to thank the NASA family and the people from around the world for their incredible outpouring of love and support. … Although we grieve deeply, as do the families of the *Apollo 1* and *Challenger* before us, the bold exploration of space must go on. Once the root cause of this tragedy is found and corrected, the legacy of *Columbia* must carry on for the benefit of our children and yours."

While the Space Shuttle *Columbia* carried no private citizens, the loss of its crew again underscored the risk of spaceflight for any humans. Today, the Space Shuttle fleet is operated by a commercial joint venture between The Boeing Company and the Lockheed Martin Corporation called the United Space Alliance (USA). The fleet's remaining three vehicles are the *Discovery*, *Atlantis*, and *Endeavour* (the replacement shuttle for *Challenger*). The Space Shuttle *Enterprise* was a test vehicle designed only for atmospheric flights and landings. It is now on display at the National Air and Space Museum's new Udvar-Hazy Center near Dulles airport outside Washington, D.C.

NASA projects flying this fleet at least through the year 2010 when, according to the new national vision for space exploration presented by President George W. Bush in January 2004, the vehicles will begin to be retired. During the next few years NASA is working to develop a new Crew Exploration Vehicle (CEV) to support its mandate for travel beyond LEO. Until that vehicle is ready, transport to the ISS will most likely be contracted to the Russian *Soyuz* vehicles.

While the Space Shuttle still flies, the opportunity for us as developers of the space tourism industry is to continue to vigorously advocate the filling of the shuttle's eighth cabin seat, which is typically not in use, with a paying passenger.

Two Thoughts on the Big Boys

In 2002, at the height of the American and Russian space programs, only about 40 people were launched into orbit. Every seat to space is coveted and won through great competition. Many astronauts and cosmonauts do not like the idea of space tourism or of private citizens paying for a seat. Several of my astronaut friends, both active and retired, have told me they feel the concept of space tourism detracts from the respected position astronauts hold in society. This perspective is quite understandable because astronauts work so hard and long to earn their seats, but it does not contribute to the ultimate goal of humans exploring and colonizing off world. I like to tell the active astronauts that once they retire from the astronaut corps, they will be in great demand from the growing private space tourism industry to fly as pilots, builders, and operators of our orbital ports, space yachts, and space cruise ships. How amazing it is to see their perspective change as they think about these potential future applications of their skills — new roles that they could play and the "big bucks" they could earn. Just because millions of people fly in commercial airliners every year does not take away the mystique associated with the word "pilot." The same will certainly hold true of "astronaut."

NASA is composed of talented people — most of whom are dedicated to the scientific exploration of space. It is a federal agency that must maintain a clear separation between its programs and private enterprise. We should not look to NASA to be a significant player in developing the space tourism industry. They can be of great assistance in technical research, and their training of astronauts will supply us with skilled and experienced personnel to assemble and operate our facilities and ships off world.

What about the United States' aerospace industry? What role will it play? The big players include Boeing, Lockheed Martin, and Northrop Grumman. In reality, the majority of these companies' revenues and profits are generated through military and aviation contracts. The space part of their business is much smaller.

I believe the role for the aerospace industry in the development of a space tourism industry is that of a research, engineering, and manufacturing contractor providing expert services. Their role should not be as the developer or operator of space tourism facilities and ships. The aerospace industry will be happy to accept multibillion-dollar contracts with well-defined programs and goals from private enterprise. Just as a cruise line approaches a shipyard with a detailed program and design for a new cruise ship, we will approach the aerospace industry as our "space shipyard" with our detailed designs and specifications. They will build the parts, launch them, assemble them in Earth orbit, test the spaceship, and then hand it over to the owner when their job is done.

Visionary corporations in the cruise line and entertainment industries with the ability to attract the necessary creative talent and to amass billions of dollars in long-term financing will be the owners, developers, and operators of the space tourism industry.

The New Space Pioneer

Creativity, ideas, intelligence, passion, dedication, and pure force of will characterize the new space pioneer. The first generation of space pioneers — the politicians, scientists, engineers, and astronauts who put men on the Moon and brought them safely back to Earth — had these same traits. Today, important new entrepreneurial skills are being added, such as writing solid business and financing plans; developing public speaking, promotion, and media relations skills; as well as combining financing with solid science and engineering.

The new space companies are beginning to establish areas of business focus such as new reusable vehicles, in-space services to the International Space Station, new launch complexes, lunar development, robotic and telepresence services, and even private space exploration. Paralleling the development of real space business ventures is the creation of the space experience simulation industry. Zero gravity aircraft flights; space camps for kids and adults; totally immersive multi-day vacations aboard full-scale mock-ups of orbital super yachts, cruise ships, and racing ships; lunar bases; Mars resorts and spas; and other space adventures will excite the media and the public — building even greater demand for the real experience.

Renaissance players in the realm of space experience include a variety of nonprofit foundations and societies that for years have been doing the political work, research, and education in service of space exploration and now the space tourism movement.

The emergence and ongoing development of several of these organizations are legitimizing the concept of public access to space travel as a viable and profitable use of space. As you consider the role you could play in this process — designer, venture capitalist, angel investor, engineer, promoter, teacher — know that your interest is shared by an increasing number of credible professionals and national organizations.

For example, in June 2000, the U.S. Chamber of Commerce announced the launch of a "broad-based effort to advance the interests of United States businesses in the commercial development of space, through the first ever Space Enterprise Council." The Chamber is the world's largest business federation, representing more than three million businesses and organizations. To date the Space Enterprise Council has conducted several workshops and symposia to introduce the potential of space development to U.S. business owners.

The Space Frontier Foundation (SFF), a nonprofit organization advocating space commercialization, began a series of Space Enterprise Symposia in 2000 designed to connect space entrepreneurs with qualified investors. These symposia have been a great success and a very important service to the entire space community. The symposium held at the SFF 2001 annual conference chaired by Rick Citron of the business law firm Citron and Deutsch (*www.candlaw.com*) attracted more than 50 business plans for new space service companies, with 10 chosen for presentation to investors. Some of the presenters did receive seed funding based on their presentations at the conference. Most of these new companies did not exist in 1999.

Below are examples of new space companies created by new space pioneers, chosen because they either take people to space or bring space down to Earth through simulated space environments and experiences.

The Spacehab Corporation (*www.spacehab.com*)

One of the real heroes of the space renaissance, Spacehab was the first commercial company to provide services to the International Space Station through its Spacehab module, which sits in the cargo bay of the Space Shuttle. The module is pressurized and connected to the crew cabin by a tunnel. It doubles the astronauts' living and working area and provides a critical storage area for supplies and equipment for the station. The company was founded in the early 1980s by Bob Citron and Tom Taylor, neither of whom had an aerospace background.

Prior to focusing on the Spacehab module, Bob had been working on space tourism concepts, including a tourist module that took up the entire cargo bay of the Space Shuttle. Tom was the creator of the Space Shuttle external tank torus (round) space station back in 1979. He and I led a diverse team for a year-long study that provided designs for the interiors of his space station for 200 people, including 30 space tourists. We published our study in 1980.

Bob and Tom devoted years of their lives to developing Spacehab, Inc., and they succeeded. Bob's brother, Rick Citron, was a key player in acquiring early private investors. Together with Shelley A. Harrison, Ph.D., they raised more than $100 million in investments and built the company until they had succeeded in flying the first Spacehab module in June 1993 aboard the Space Shuttle *Endeavour*. In late 1995, Spacehab went public. They now manufacture and provide a double module that fills almost half of the shuttle cargo bay. Spacehab modules had flown almost 20 times by the end of 2002. Spacehab, Inc., is the first great example of private citizens conceiving a space project and implementing it. What space project is on your drawing board?

Universal Space Lines (*www.universalspacelines.com*)
Many of us marked the beginning of the transition from total government domination of access to orbit to opening the door to private enterprise with the successful first test launch of the unmanned Delta Clipper-Experimental (DC-X) on August 18, 1993. The DC-X was the world's first reusable vertical-takeoff-and-landing (VTOL) rocket. The pioneers' long-term goal was to create a wingless Single-Stage-To-Orbit (SSTO) vehicle. The DC-X program was funded by the Strategic Defense Initiative Organization (SDIO) at $60 million — a small price for such an innovative and trailblazing technology program. McDonnell Douglas was chosen as the prime contractor. It is still amazing to watch video of the DC-X taking off, rising several thousand feet to stop in mid-air, then move horizontally, stop, move back, and then slowly descend to its launch site for a safe landing.

Unfortunately for everyone working to achieve low-cost access to space, the promising DC-X crashed on its final test flight. But the proof of concept had been achieved. The DC-X program proved a rocket system could have airline-style maintenance and operation with a ground support team of only 12 technicians and a small trailer for the mission control center. In the long term, VTOL is the only approach that would allow a return to the Moon with a vertical-landing rocket just like the successful *Apollo* Lunar Module. Another advantage of this kind of approach is its ability to provide hypersonic suborbital flights. For example, you could lift off from the Los Angeles International Air/Spaceport and land at Tokyo International Air/Spaceport within an hour, creating a potential market for the military, business passengers, high-value packages, and eventually the tourists who want to experience several minutes of zero gravity.

After the completion of the DC-X program in the mid-1990s, the directors of the program, Dr. Bill Gaubatz and *Apollo 12* commander, Astronaut Pete Conrad, formed the independent private rocket development company, Universal Space Lines. The use of the word "lines" was a clear indication of their goal to develop a "space line" based on the airline and cruise line models. One of Pete Conrad's favorite sayings was, "Our motto is: Anywhere on the planet in 45 minutes or less."

Space Adventures, Ltd. (*www.spaceadventures.com*)
"Making private spaceflight a reality today" is the statement of Space Adventures, the world's premier space tourism and space experiences company.

Space Adventures has a distinguished advisory board of astronauts, cosmonauts, scientists, and other leaders in the space tourism movement. Many of these world-class experts go on tours of Earth-based space facilities, providing lectures and personal insights.

The company has created a wide variety of exciting programs and activities all designed to connect people to space. Through collateral materials and online at their website, Space Adventures outlines their *Space Flight* and *Steps to Space* programs.

In 2001, Space Adventures enabled Californian businessman Dennis Tito, to realize his dream of spaceflight and then, in 2002, South African Internet tycoon Mark Shuttleworth, became the second private space explorer. In 2004, Gregory Olsen, Ph.D., went into training at the Yuri Gagarin Cosmonaut Training Center in Star City, Russia, and will fly into space as the third such client. He will focus on scientific research during his mission on the International Space Station. Each year, Space Adventures makes remarkable advances in opening the final frontier to private citizens and they are continually working with numerous other wealthy individuals who want to have the experience of real spaceflight. Eric Anderson is their president and CEO.

More than 100 "Citizen Explorers" have made $4,000 deposits with Space Adventures to reserve their seats aboard private spacecraft and aerospaceplanes now in development. These vehicles could provide suborbital flights with up to 15 minutes of zero gravity. The total cost of the week-long training, flight, and post-flight activities is $98,000. Space Adventures will fly tens of thousands of people in space over the next 10 to 15 years and beyond, both orbital and suborbital, around the Moon and back, from spaceports both on Earth and in space, to and from private space stations, and aboard dozens of different vehicles. Their plan is to continually provide newly available space experiences and improve existing space experiences.

LunaCorp, Inc. (*www.lunacorp.com*)
This Arlington, Virginia-based company was founded by David Gump in the early 1990s. He has pioneered a concept to send a privately financed small rover to the Moon that would send back video and data from the lunar surface. It would also provide one of the first applications of telepresence. In 2000, Gump announced support and funding from Radio Shack, a real accomplishment for our nascent simulation industry. Having created the concept of a "tele-tourist," LunaCorp's goal is to make it possible for Earth-bound humans to drive the company's lunar rovers robotically, while they sit in simulators on motion-based platforms, feeling the motion of the rover over moon dunes and across lunar craters.

The Zero Gravity Corporation (*www.zerogcorp.com*)
This company is the first in the United States to offer rides to the general public on aircraft specially equipped to produce 30 seconds of zero gravity through parabolic flights. Modeled after the NASA zero gravity aircraft flight program started in the 1960s, this private company will provide a full-day total experience for the public and for corporations. They will also provide services to research scientists and the entertainment industry for filming in zero gravity.

Bigelow Aerospace (*www.BigelowAerospace.com*)
Founded by Robert Bigelow, owner of Budget Suites hotels and apartments based in Nevada, Bigelow Aerospace is researching and designing an inflatable space station modeled around the International Space Station with the intention of significantly reducing development and operation cost of orbital facilities. They plan to streamline the entire design-through-manufacturing process, and play a critical role in the establishment of the space tourism industry.

There are several other private enterprise ventures whose website addresses are noted in the resource section at the back of this book. Following are some of the nonprofit educational and research foundations and societies that are pioneering their areas of expertise for the benefit of the space renaissance.

U.S. Space Camp (*www.spacecamp.com*)
This famous space training mecca is based in Huntsville, Alabama. From 1982 through the end of 2002 they have graduated more than 400,000 students. Today, more than 15,000 kids each year in spaceflight uniforms spend five days at camp learning to be astronauts. Graduation consists of flying simulated Space Shuttle and International Space Station missions in high-quality, full-scale mockups. The Huntsville facility also provides an exciting and usually sold-out adult Space Camp program. Almost 250,000 day-visitors tour the Space Camp grounds to watch the cadets train and perform their missions, explore the space museum, and walk through the world's largest rocket park featuring more than a dozen real rockets, including the huge, 363-foot tall *Saturn V* rocket that launched *Apollo* astronauts to the Moon. There are approximately a half dozen space camp-like programs and facilities around the world.

The Challenger Learning Center (*www.challenger.org*)
This nonprofit educational foundation was established by the families of the crew lost when the Space Shuttle *Challenger* disintegrated in 1986. There are more than 50 centers in operation that provide more than 300,000 children and teachers each year with team learning experiences using space-themed simulations. The goal of the centers is to foster long-term interest in math, science, and technology in students, and to inspire them to pursue careers in those fields. New programs are constantly being added, including students constructing their own simulated "Mars Colony." Challenger Learning Centers online have the potential to serve millions of children around the world. The Challenger Center has become a living memorial to all space explorers and a great success story.

The Space Tourism Society (STS) (*www.spacetourismsociety.org*)
 founded STS in 1995 with the main goal: "To stimulate a profitable and expanding space tourism industry. To conduct research, build public desire, and acquire the financial and political power to make space tourism available to as many people as possible as soon as possible."

STS is focused on modeling this new industry after the highly successful super yacht and cruise line industry. Exploring the beautiful, sensual, and futuristic aspects of space tourism is essential if we are to gain the interest and support of the world public and the finance community. STS is modeled after the National Geographic Society in that it conducts real research into several technical, social, design, and marketing areas and then promotes its findings to the public.

As of mid-2004, it is the only nonprofit professional society in operation whose sole focus is the development of the space tourism movement and industry. We hope other similar groups will spring up around the world.

The Ansari X Prize Foundation (*www.xprize.org*)

This St. Louis-based nonprofit organization is offering a $10 million prize to the first private group to privately finance, build, and successfully fly a reusable vehicle to at least 100 kilometers (62 miles) away from Earth (the official recognized beginning of outer space). The vehicle must be capable of carrying three people and must repeat the flight in the same vehicle within two weeks. This is called a suborbital flight because the vehicle will not have the speed or energy to enter Earth orbit. It is the same type of parabolic flight that the first American in space, *Mercury* astronaut Alan Shepard, flew on May 5, 1961.

The Ansari X Prize mandates three passengers in order to force contestants to consider the requirements of passengers during flight. The X Prize is modeled after the Orteig Prize, won by Charles Lindbergh in 1927 when he completed the first flight from America across the Atlantic Ocean to Europe. In those days, the aviation industry successfully used a series of cash prizes to stimulate innovation and public awareness. To date there are more than 25 teams registered as official contestants for the Ansari X Prize competition.

Originally called simply the "X Prize," the name was changed on May 5, 2004 (the 43rd anniversary of Alan Shepard's *Mercury* flight), to the "Ansari X Prize" in honor of two Iranian-born entrepreneurs, Anousheh Ansari and his brother-in-law, Amir. They provided a multimillion-dollar donation to the organization for continuing support of private space travel. Their participation also shows the international flavor of space tourism. "As a child I looked at the stars and dreamed of being able to travel into space," said Anousheh, an avid space enthusiast. "As an adult, I understand that the only way this dream will become a reality is with the participation of private industry and the creative passion of smart entrepreneurs. The Ansari X Prize provides the perfect vehicle to ignite the imagination and passion of fellow entrepreneurs, giving them and their courageous pilots a platform for success."

A Pioneering Age

The Space Age formally began on October 4, 1957, when the Soviet Union launched the first artificial satellite, *Sputnik 1*, into Earth orbit. People in the United States listened to their radios with wonder and some fear as the basketball-sized sphere orbited the Earth, sending back blips and bleeps. Then, on April 12, 1961, Yuri Gagarin, a Soviet cosmonaut, became the first human to orbit the Earth in the *Vostok 1* capsule. John Glenn launched into Earth orbit on February 20, 1962, and the Space Race was officially on. The Space Age i characterized by grand words and grand deeds, including President John F. Kennedy's historic Moon speech. The *Mercury, Gemini,* and *Apollo* programs. The Moon landings.

The term "Space Age" was first popularized in the early 1960s. It signaled entry into a new age of utilizing outer space. We were traveling into orbit on a regular and accelerating basis. But the Space Age also meant something much more important. Our human horizon had expanded and so had our aspirations and expectations. We had crafted new dreams and created new challenges toward which to strive.

While the technical and heroic achievements of space travel since then have been truly astounding, that expansion of our horizon and aspirations has been the greatest accomplishment. Humanity now has an unlimited frontier into which we can explore and expand, a frontier that will always provide us with surprises, challenges, and opportunities.

Humanity can be proud of its space accomplishments since the beginning of the Space Age: The awe-inspiring sight of the first human-witnessed Earthrise from behind the Moon; the footprints left on the lunar surface and the plaque on a leg of the lunar lander, reading, "We came in peace for all mankind;" the first view of the red Martian landscape from eye level by *Viking*; the grand tour of the Solar System by the *Voyager* probes and the knowledge that they will someday reach the stars; the first view of the *Eagle* Nebula, a "birth place of stars," from the *Hubble* Space Telescope; the Space Shuttle fleet making its 100th flight on October 11, 2000, with the flight of *Discovery*; the fact that women and minorities have earned their rightful place as equal explorers of the space frontier; the sight of the International Space Station crossing the night sky, easily visible to anyone on the Earth who wants to look up.

At the beginning of this new century, just over 400 people have ascended into Earth orbit, some up to seven times. Twenty-six men have orbited the Moon and 12 have landed on it. A few cosmonauts have lived in orbit for more than one year on a single mission. With few exceptions, all of these explorers have been members of a military or government space program. Today, this exclusive club is beginning to open to private enterprise and private citizens *are* going into orbit.

In my opinion, the year 2000 represented the end of the Space Age. The history books can now include a new entry. The stage is set for a new age of space development. I call this new age the "Orbital Age." The Orbital Age will be set apart by its atmosphere of inclusion rather than exclusion. It will exemplify the well-recognized megatrends of decentralized government, which will attract new players, new ideas, large-scale diversified financing, and stimulate a direct connection between space ventures and the world public. Orbital and space tourism will grow to be the largest off-world business, further connecting people to life and opportunity beyond Earth.

The Orbital Age will expand, diversify, and prosper through the year 2050. Its milestones include orbital tourism; a return to the Moon for the beginnings of lunar tourism and industry; the first human exploration and colonization of Mars; utilization of the inner planets, Venus and Mercury; and a continuation and expansion of our robotic exploration of our Solar System and beyond.

Around the year 2050, we should be ready to close the history books on the Orbital Age and open a new one called the "Solar System Age." Humans will venture to and explore the outer giant planets, including Jupiter and Saturn. During this age, we will build huge space colonies, terraform Mars, and populate the Solar System with millions of people. The limitless resources and knowledge gained will enable and inspire us to clean, preserve, and protect Earth.

By the year 2100, we may be opening the next stage of human evolution, which may be called the "Stellar Age," as we humans follow our robotic explorers to the nearest stars. We may begin to physically and socially evolve into a true Solar System Species.

Conclusion

We are making real progress in breaking the bonds of gravity and the old perceptions that only governments and large aerospace corporations do real space ventures. The United States has a new vision for space exploration. And heroic individuals are emerging to build successful space companies and embark on space travel.

Chapter One sets the scene for the current state of space travel. In it I introduce the business model that I believe best supports and makes achievable the creation of a space tourism industry.

The very fact that you are reading this book shows that you are interested, intrigued, and inherently aware of the potential of private space enterprise and, particularly, space tourism. The best news yet — you can join this important movement today. Your ideas, talents, and energy can be woven into the shimmering fabric that is the space renaissance.

PART ONE
SPACE TOURISM

CHAPTER 1
SPACE TOURISM

"Every revolutionary idea … seems to evoke three stages of reaction: They may be summed up by the phrases: (1) It's completely impossible — don't waste my time; (2) It's possible, but it's not worth doing; (3) I said it was a good idea all along."

Sir Arthur C. Clarke in The Promise of Space, *September 1967*

Introduction

For the space renaissance to enjoy long-term success, it requires unlimited market potential and a connection to the world public. Together these elements will inspire the development and operation of a multitude of space-based profit-making businesses.

The key driver behind those elements must be space tourism — the true hero of the Orbital Age. It is the only kind of space-based business for which there is an existing market base of millions of people — millions of people who want to go.

Space tourism directly connects people with space. Space tourism also has huge prestige and profit-making potential, and it clearly attracts the interest of the media. It also attracts the interest of women, who are our most important market opportunity and challenge.

Those of us leading the space tourism movement are creating a new market niche in the ultra-high-end exotic tourism business. Our "exotic locale" is actually much closer since LEO is only 100 miles from anywhere on Earth. Our primary obstacle to reaching that location is cost. Currently, costs for manufacturing and launching craft to space — craft that, with the exception of the Space Shuttle, never get used again — is prohibitively expensive for the majority of citizens. And, given the number of people who can participate in space tourism today, space is relatively dangerous to reach and to return.

However, over time, and as market demand grows, technology and safety will dramatically improve. Costs will come down, making Earth orbit accessible to a larger market. It will happen.

The evolution of travel to unique locations has always followed this pattern. Think about it. First there are the pioneers — typically government or corporate travelers — whose task is survival and to conduct research, but not necessarily to have fun. Following the pioneers comes the exclusive group of adventure travelers, stout of heart and deep of pocket. Eventually, that market expands and matures, followed by a mass market. This pattern has brought humankind to formerly inaccessible and dangerous locations in growing numbers. Mt. Everest, the African plains, Antarctica, and the rain forests, are but a few.

The Why

Why create a space tourism industry? To provide people with truly unique experiences they can have only off world.

This is the most important point in this book. Pioneers standing on the edge of the frontier of space tourism must understand and remember we are in the experience business. Not the space business and certainly not the launch business. Facilitating an individual's experience of the unique qualities of space travel is our most important concern. All other technologies and issues must exist only to serve this overriding goal.

The space tourism experience must be luxurious, sensual, romantic, spiritual, inspiring, exciting, and fun if it is to attract the pioneers and the adventure travelers and ultimately millions of people. The quality and variety of the food, recreation, personal attention, and services must evolve over time to rival the best on Earth if we are to achieve high passenger satisfaction and return bookings.

The What

What do people want who are interested in going off world? I have asked this question of people from around the world. Their answers echo those revealed in several recent professional market surveys:

In order of interest, people want to:
- Experience the freedom to float, fly, play, dance, compete, and make love in zero gravity.
- Experience the beautiful views of Earth, the Moon, and space — to see their countries and cities from Earth orbit.
- Experience a sunrise or sunset every 45 minutes.
- Take photos and video of their experiences.
- Listen to their favorite music while looking out the viewports.
- Have a life-changing experience that offers the opportunity for greater comprehension of the universe and themselves.
- Share these special experiences with people of similar interests and passions.
- Contribute to the expansion of the human species into limitless new opportunities and exciting challenges.
- Have the right to brag about having gone.

The Who

I have always been amazed by the wide diversity of people interested in going to space. People of every age group, race, religion, political system, and economic status have expressed excitement over the prospect that they, or a family member, may go someday. Space travel and exploration are truly of universal interest and, to many, a lifelong passion.

In the long run, this universal interest may well be one of the most important aspects of becoming a spacefaring species. That we can travel to space as a people of Earth, and not of competing nations, is extremely appealing.

Space tourism industry pioneers are already of very diverse backgrounds and are even sometimes surprised we are in the space tourism business at all. We began our careers as hard-core space exploration and colonization advocates who learned over the years that there would never be enough government funding or even public interest in scientific or industrial applications of space. It is just not sexy enough!

We learned we needed to break the government's hold on access to space and to privatize the entire process. While working to identify other space-based industries worth such a level of investment, we consistently returned to space tourism. As I have stated before, only a successful private space tourism industry can drive the development of technology and infrastructure to a level that can affect the long-term prosperity and expansion of the human species.

The How

In 1982, with all these possibilities in mind, I created a clear course of design and development for building the space tourism industry. My approach is to model the space tourism industry after the cruise line industry.

It would have seemed natural to assume the space tourism industry would be modeled after the airline industry. In fact, the aerospace industry and even NASA are quite comfortable with that assumption. However, the airline industry is only in the transportation business, not in the hospitality or experience business. Their main concerns are safety and cost reductions, and moving the most people in the quickest and most efficient manner.

In comparison, cruise lines are in the hospitality and experience business. They are focused on passenger satisfaction, providing high-quality overnight accommodations, great food, and entertainment for anywhere from three days to two weeks. They are our kindred spirits — their goal is to create unique and memorable experiences for their passengers.

The cruise lines take millions of people safely into one of the harshest environments on Earth — the ocean (many say the ocean is a more harsh environment than space) — using increasingly complex technological marvels of ships. And they make a profit. They have excellent passenger satisfaction and repeat bookings, and have succeeded in creating a worldwide culture around the cruise experience. The cruise lines are the perfect model for the space tourism industry. I have convinced rocket scientists, aerospace engineers, and even astronauts that this is the correct industry model. I will expand on this model later in this chapter.

Surviving the Giggle Factor

People used to chuckle about the idea of space tourism. But the concept of real space tourism for private citizens has gained significant momentum since 1996. In that year, The X Prize, founded by Dr. Peter Diamandis, was announced and endorsed by the head of NASA. The Space Transportation Association (STA) and NASA cosponsored the first professional workshop focused on defining and creating a space tourism industry, held at NASA Headquarters. This milestone workshop was cochaired by Tom Rogers of STA and Ivan Becky of NASA.

One year earlier I had founded the Space Tourism Society (STS), which began holding its first public meetings and generating positive articles and news reports on space tourism. Dr. Patrick Collins, a respected economic academician, published papers and articles on surveys he had designed and conducted in Japan on space tourism (*www.spacefuture.com*). In an article for the May/June issue of *Innovations* magazine, John C. Mankins, a senior NASA official in the department of advanced concepts at NASA Headquarters, wrote, "Space tourism will come. It is as inevitable as the Panama Canal and as irresistible as the development of communications satellites."

Before 1996, media mocked the idea of private citizens going into space. We heard the giggles. We endured them from media, NASA, aerospace officials, and even colleagues in the space movement. But we persevered. Today, as I wrote earlier, people are "getting it." Today, the media ask about the business, finance, marketing, and new technology potential of space tourism. They ask to identify leaders in the exciting new space tourism movement. Today, many of those who once giggled at our wild idea of space tourism are scrambling to get on board.

There has been steady growth in the numbers of professional workshops, international symposia, conferences, expos, and promotional contests focused on space tourism. People are now writing books (like this one). They are authoring dozens of articles a year for respected newspapers such as *The Wall Street Journal*, *The Los Angeles Times*, *The New York Times*; and for magazines such as *Fortune*, *Scientific American*, *Wired*, *Popular Science*, *Harper's Bazaar*, *Time*, and *Newsweek*. There have been several television programs, including NBC's *The Today Show*, and major networks including The Discovery Channel, PBS, and ABC News, that have developed features and programming on space tourism. We were amazed and delighted with the massive positive international media coverage of Dennis Tito's flight.

Dozens of new space tourism companies have been formed by billionaires, retired astronauts, and inspired entrepreneurs to finance, design, and build the new industry. Millions of dollars each year are being invested to establish identities, names, and claim business territories. Some of these new space tourism companies include Universal Space Lines, Space Adventures, Red Planet Ventures (RPV), LunaCorp, the Zero Gravity Corporation, MirCorp, Bristol Spaceplanes, Virgin Galactic Airlines (started by English billionaire and adventure capitalist Richard Branson), Space Tours, Orbital Properties, Pioneer Rocketplane, and others.

Nonprofit societies and foundations such as the Space Tourism Society, the ShareSpace Foundation (founded by *Apollo 11* astronaut Buzz Aldrin, Ph.D.), the Ansari X Prize, the Space Tourism Club, Spacefuture.com, and others have been formed to research and promote space tourism.

Older general space support groups such as the Space Frontier Foundation (SFF), the National Space Society (NSS), the Space Foundation, the Challenger Learning Centers, the Space Studies Institute (SSI), and others are now comfortable discussing and writing about the subject of space tourism. Some have incorporated the subject into their conferences and programs. Professional aerospace organizations such as the prestigious American Institute of Aeronautics and Astronautics (AIAA), and the American Astronautical Society (AAS) now request papers on the subject of space tourism for their conferences and publications.

The International Space University (ISU) was founded in the early 1980s by students at MIT, including Peter Diamandis, who later founded the X Prize and the Zero Gravity Company. ISU and other respected universities such as George Washington University, MIT, Carnegie Mellon (Robotics Institute), Rochester Institute of Technology (School of Food, Hotel, and Travel Management), and others have recently produced space tourism studies. Some are now offering classes in space tourism research, planning, design, and management.

And even the government has finally taken notice. Not long before this book went to press, the U.S. House of Representatives had passed H.R. 3752, "The Commercial Space Launch Amendments Act of 2004," which is considered a milestone in supporting the efforts of space entrepreneurs to invest in new rockets and to fly those vehicles. If passed by the Senate and signed into law, the bill will give regulatory authority over human flight to the Federal Aviation Administration's Office of Commercial Space Transportation. This important goal finally eliminates the confusion over who should be regulating flights of suborbital rockets carrying human beings. The office will also be charged with developing regulations for crew pertaining to training and medical conditions, and to ensure that space tourists are informed of spaceflight risks. If passed, the bill will "help nurture [commercial human spaceflight] while at the same time ensuring that public health and safety are protected," according to House Science Committee member Bart Gordon.

Tourism Industry Takes Notice

The most important step forward for the space tourism movement has been the inclusion of travel and tourism industry professionals in the discussion. There were pioneering efforts by Society Expeditions in the early 1980s to develop reusable space tourism passenger vehicles and to book passengers. This effort was headed by Bob Citron. Gloria Bohan entered the space tourism field in 1996 through Tom Rogers and STA. Gloria is the founder and president of Omega World Travel, a company she and her husband have built into one of the largest privately-owned travel agencies in the world. Omega books more than $700 million in travel each year.

In 1998, she, Mike McDowell, Peter Diamandis, and Eric Anderson cofounded Space Adventures, Inc., becoming the first professional space travel agency focusing on space-themed experiences both on Earth and off world. McDowell is the founder of Quark Expeditions, one of the premiere adventure travel groups. Since 1989 Quark has assisted more than 30,000 people to visit the Arctic and Antarctic. The company is run by Eric Anderson and his growing team of space tourism pioneers. Through its partnership with the Russian Federal Space Agency, formerly Rosaviakosmos, Space Adventures has assisted private citizens in achieving their dreams of spaceflight by helping them secure seats aboard *Soyuz* taxi flights to the International Space Station. Dennis Tito became the first space tourist in April 2001. In April 2002, Mark Shuttleworth became the first South African in space. Gregory Olsen will follow as the third space tourist, probably in 2005.

The January 2000 issue of the American Society of Travel Agents (ASTA) magazine featured the first article on space tourism from a major travel association. It was written by Judy Jacobs and titled "The Dawn of Space Tourism." It was a comprehensive overview of all the key players, the status of the new industry, and plans for the future. The cover of the magazine was a beautiful image of Earth seen from space through an airplane-style window. This article reached the majority of the travel agents in the United States and provoked a series of other articles, television specials, and speeches at travel conferences.

Robert L. Haltermann is the former executive director of the STA's Space Travel and Tourism division, and remains a member of the STS's Board of Governors. He has extensive experience in the travel industry and with cruise lines. In 1996, Bob wrote an excellent report titled "Evolution of the Modern Cruise Trade and Its Applications to Space Tourism" in which he skillfully outlined the similarities between the ocean-going cruise experience and possible orbital cruise experiences. He also noted how the travel and tourism industry and travel agencies could be key players in creating and selling the space tourism industry.

In response to this rising interest, I have spoken at several travel and tourism conferences including ASTA in early 2000, and the RSA Summit Conference for the International Travel Industry in 2001. Many of these top travel professionals are now excited about space tourism becoming a new segment of the tourism industry.

The adventure travel industry is paying closer attention to space tourism via the Adventure Travel Society, founded by Jerry Mallett, and through several companies who specialize in taking affluent travelers to exotic places such as Antarctica, Easter Island, or in deep sea diving vehicles to see the *Titanic*.

In 2000, British Airways produced and broadcast a series of beautiful television commercials showing advanced aerospaceplanes with their logo-emblazoned bodies transporting passengers to and from space holidays. The campaign proved to be one of their most successful.

In 1998, my team produced a 40,000 square foot Space Fair, which was held aboard the *Queen Mary* cruise ship in Long Beach, California. With "Come Celebrate the Dawn of Space Tourism" as its theme, the fair featured Apple Computer, Space Adventures, Boeing, Lockheed Martin, Universal Studios, HBO, Warner Brothers Studios, Sprint, Kodak, the Challenger Foundation, the Space Tourism Society, the Orange County Space Society, and dozens of other organizations interested in associating their products and services with space tourism. Space Fair 98 attracted 15,000 visitors and generated more than 20 million positive media impressions on the new concept of space tourism.

The success of these public space tourism shows clearly indicates that this new industry is beginning to establish itself and that it can attract the attention of well-established industries, the media, and the public.

In the early 1980s there were only a half dozen knowledgeable professionals advocating the establishment of a space tourism industry as the financial key to opening up off-world commerce and exploration. As of 2004, there are hundreds of professionals and more than a dozen dedicated companies and societies around the world taking up the quest for space tourism.

We have succeeded in creating and sustaining the space tourism movement.

The Cruise Line Model

Several people who enjoy ocean cruising have told me that, on a moonless night far from land with the sea calm and stars blazing from horizon to horizon, they felt as though they were cruising in outer space. All have said they would love to take a cruise in Earth orbit someday.

Some people love ships and cruising. There is something magical about boarding a beautiful ship, then the celebration to leave port to begin a relaxing and fun voyage. Cruising is all about rewarding yourself, forgetting about your stressful life and giving yourself permission to relax. It is about giving yourself permission to open up to new experiences, people, and places. For millions of people around the world cruising has become a lifestyle and a culture. Some take at least one cruise a year on a favorite line and even a favorite ship. Some cruisers who spend weeks aboard ship each year can easily accumulate a year or more at sea during a lifetime of cruising.

Onboard a modern cruise ship you are safe, cared for, entertained, fed, and kept the center of attention. From the captain to the housekeeping staff, the crew all know they make their livings by providing you with a satisfying vacation experience that brings you back.

The cruise line industry has matured during the past 200 years. There are close to 50 cruise lines around the world taking almost 12 million people each year out for cruise vacations. The lines range in size from the largest — Carnival Cruise Lines with almost 40 large ships — some accommodating up to 3,000 passengers and 1,500 crew members — to the smaller, intimate cruise lines with two to three small ships catering to less than 100 exclusive passengers plus crew.

The owners of the cruise lines do not consider themselves in the shipping business, even though they own and operate ships. They clearly consider themselves in the hospitality and leisure business. Their facilities just happen to be ships at sea.

This industry is one of the most international of businesses. The lines are owned by corporations in Britain, the United States, France, Italy, and Asia. Ships are designed in the United States and Europe and then built in Europe. Most of the command and engineering staff are English, the leading chefs are French, the stewards and hospitality staff are Italian, with the ship's service personnel such as waiters and maids being from the Philippines or Mexico. Most of the entertainment staff is from the United States. One of the best books on the history and evolution of the cruise line industry is *The Liners — A Voyage of Discovery* by Bob McAuley, with consultant William H. Miller, published in 1997 by Motorbooks International.

Since the mid-1980s there has been a boom in the cruise line industry. The age range of the cruising public has dramatically expanded to include a much younger audience. As middle classes grow in number around the world, more people are seeing cruising as an interesting and affordable vacation. There have been billions of dollars invested in building dozens of new ships and in modernizing older ones.

Like any mature industry, the main players have specialized in specific market segments. The high-end lines such as Crystal Cruise Lines focus on the wealthy senior citizen with ships of 600 passengers taking longer cruises, some lasting for months at a time visiting dozens of high-end ports. Their liners carry mature names such as *Harmony of the*

Sea or *Splendor of the Sea*. The Princess Cruise Line caters to the middle-age-cruisers, many taking their children. Their ships have names such as *Grand Princess*, *Royal Princess*, and *Sun Princess*. A mid-sized ship carries about 1,500 passengers. Carnival Cruise Lines specializes in the new market of young cruisers. Their "Fun Ships" carry names such as *Mardi Gras*, *Festival*, and *Carnival* — all inspiring the party theme. They build the world's largest cruise ships with the goal of providing enough activities and services onboard so that passengers, and their money, stay onboard.

The space cruise lines of the future will operate in a similar manner to the ocean cruise lines. Most people actually fly to an ocean port before boarding their ocean cruise ship at a cruise line terminal. Space tourists will fly into orbit aboard a variety of orbital access vehicles, dock with an orbital port, and then board their orbital or lunar cruise ship.

I forecast that the ocean cruise lines will extend their passenger services first into the skies with the reemergence of the huge airships and "sky cruising," then eventually into orbit and beyond. The off-world lines will specialize in specific market segments based on passenger preference.

Building a cruise ship starts with extensive market studies and trend analysis. Cruise lines focus on specific market segments and design ships to cater to the desires and needs of those segments. They then contract with a shipyard to build their ship and outfit it. After trial runs, the ship is handed over to the cruise line and the shipyard is no longer involved. More testing and crew training follow, then the maiden voyage. Most ships will serve for at least 30 years before being sold or scrapped. A ship will have two dry dock stays of about six months during its operational life. The lines commission at least two and sometimes four ships of the same class to reduce the construction cost of each ship.

The cruise lines and the U.S. Navy have established officer training academies. We will someday need a real space academy that can translate the logistical support systems and experience to orbital operations.

Shipyards are using more robotics to build the cruise ships. The ships are becoming more automated, reducing the engineering staffs, allowing more crew members to focus on direct passenger interaction, entertainment, and services. This trend of using more robotics and automation in ship-building and operations will be critical for economic operation of off-world ships and facilities on the Moon

I envision the aerospace industry playing a role similar to ocean shipyards. After companies manufacture parts of spaceships and orbital or lunar facilities, then they are delivered to orbit for assembly and testing, and everything is transferred to the owners and operators. The work of the aerospace company is done, ending the days of multiyear or multidecade maintenance and servicing contracts.

The Vision Takes Shape

With a foundation of increasing professional, corporate, and public interest, we can legitimately plot a course for development of this industry. Realistically speaking, it will take decades of dedicated and smart work to evolve the space tourism industry to a level where tens of thousands of people are going off world for vacations and sporting events each year. But there is a lot of fun we can have along the way.

As just mentioned, our model for space tourism defines an industry evolving as did the cruise lines, with a parallel track modeled after the ocean-going super yacht industry. Super yachts are small, privately-owned, funded, and operated cousins of cruise liners. Orbital super yachts will function in the same manner as their Earth-bound forebears. They will be owned by the world's richest companies and individuals who will use them as status symbols, as well as for fun and rewards. These beautiful space yachts will also be used as political, sponsorship, and marketing tools.

The orbital super yachts will precede the orbital cruise ships because the yachts will not have to be designed, constructed, and operated to make a profit. They will be assembled in orbit using the most advanced technology and artistic designs available. They will accommodate between six and 20 passengers in ultra-luxurious accommodations. Orbital super yacht clubs will be their home base. Mirroring exclusive yacht clubs on Earth, the orbital yacht clubs will be funded by club member fees. Clubs will sponsor orbital super yacht races just as they do on Earth.

The technology and operations management skills gained through the development of the orbital super yacht industry will spin off to develop the more mainstream orbital cruise line industry. These much larger orbital and lunar ships will be profit-oriented, charging 100 to 500 passengers for their ultimate cruise. Due to their size and need for economical operations, the cruise ships will be a series of inflated spheres — fully functional biospheres — run mainly by robots, telepresence operators, and artificial intelligence, so that the human crew will have time to focus on serving, entertaining, and interacting with the passengers.

The sports industry will also play a major role in the development of off-world tourism. They will assemble zero gravity sports stadiums in orbit and eventually on the Moon. Basketball in zero gravity will literally add a new dimension to the sport. Racing, gymnastics, dance, and even art will find entirely new means of expression in zero gravity. Someday the Olympic Games could be held off world. Dune buggy races around the entire Moon will provide incredible challenges. There will be guided tours of the historic *Apollo* landing sites, and tourists will be able to build sandcastles on the Moon.

The space simulation industry is also rapidly evolving. Soon you will be able to take two- to seven-day, totally immersive space cruises, while participating in orbital super yacht races aboard full-scale mockups of these exciting ships of the future. You will train to be and dress as a "Simnaut" to participate in highly interactive off-world "SimExperiences." For the cost of a conventional cruise vacation, you can take a two- to five-day vacation in space and, at the same time, make a contribution to the better understanding of human factor design and orbital operations.

The SimExperience industry will become our greatest marketing and promotional tool for establishing the orbital space tourism industry, providing us with the funds for long-term research in critical technologies and operations. Full-scale simulated Moon bases, orbital cruise ships, racing space yachts, a Mars resort and spa, and the opportunity to train and serve in the Space Guard Service (modeled after the U.S. Coast Guard) will provide a wide range of space adventures. Saving the passengers of a stranded space cruise ship, capturing space smugglers, or protecting the Earth from an incoming meteor will challenge and thrill the dedicated adventure vacationer.

The Experience Takes Shape

Many ocean cruising traditions will easily translate off world. Orbital cruise ships will have beautiful names, designs, colors, and logos. There will be celebrations when an orbital or lunar cruise ship embarks from an orbital or lunar port. The port and other nearby ships will flash their lights, wishing the departing ship a safe and wonderful voyage. Passengers will have selected their own stateroom, each with a magnificent view and customized features. There will be "lifepod" drills just as there are lifeboat drills today on ocean cruise ships. The human and robotic stewards will cater to small groups of passengers, learning their preferences and needs. There will be the important cruise director, head chef, and the top restaurant *maitre d'*. There will be an exercise coordinator, and many ships will have casinos.

 Ocean cruise ships offer a variety of lectures, special tours, and some sports. These easily translate to orbit or to the Moon. Some passengers enjoy snorkeling and diving when their ship is in port. Many off-world passengers will enjoy taking space walks ("floats") and walking on the Moon.

 Transferring the habits and idiosyncrasies of life onboard a ship has clearly taken hold on the International Space Station. Many of the U.S. astronauts were, or still are, in the U.S. Navy, so they naturally pass along its traditions. The first International Space Station crew brought a small ship's bell and attached it next to the main entry airlock. When a new crew member boards ISS, they must request permission to come aboard. After permission is granted, the commander or another crew member rings the bell twice to welcome the new crew member.

Marketing Parallels

It is no secret that cruise lines market to women. As key consumers, women make 70 percent of the decisions on where a family or couple will go for vacation. The cruise lines know this, and design their brochures, sales programs, and of course, their ships to attract the attention of women. You will never see a dirty dish or unmade bed in a cruise ship brochure. You will also never see a shabbily dressed or frowning customer.

 For the young, active cruisers between 25 and 40, size does matter. They want a wide variety of physical activities available onboard. Cruise lines that specialize in this growing young market are designing and building huge new ships larger than naval aircraft carriers, displacing 150,000 tons, that are too large to pass through the Panama Canal. These ships are small floating cities with casinos, shopping malls, Broadway-scale stage shows, rock climbing walls, simulation and game centers, and nine-hole miniature golf courses. Some of the cruise lines are competing to build the biggest ships in history. This represents billions of dollars in new investments to fulfill the growing demand for the "cruise experience."

 This bodes well for space tourism because I forecast that ocean cruisers and the yachting culture will be fascinated by the prospect of cruising in Earth orbit. I have had several discussions with senior-level people from both industries, and they are intrigued. The Space Tourism Society has been invited to exhibit our space cruise ship designs and to speak at travel conferences and boat shows. STS had the first pioneering article on simulated orbital cruising in the September 1999 issue of *Sea*, the West Coast's premiere magazine on yachting.

Another reason to base the space tourism industry on the cruise line model is their logistical support and operation systems. The cruise lines make money only when their ships are at sea. Therefore, dock time is debt time, so cruise lines operate one of the most efficient servicing systems in the world. A large cruise ship with 2,500 passengers docks in the morning. The passengers and their 10,000 pieces of luggage are off-loaded by noon. Hundreds of crew members leave the ship and are replaced. Hundreds of tons of food, drink, and other supplies are loaded. Hundreds of tons of trash are also off-loaded. The ship is fueled, cleaned, and maintenance is done. All of the guest rooms and public spaces are cleaned and restocked. At approximately 2:00 p.m., 2,500 new passengers begin boarding with *their* 10,000 pieces of luggage. By late afternoon the ship leaves the dock, ready for fun and profit. In less than eight hours, 5,000 passengers, 20,000 pieces of luggage, 1,500 crew and 200 dock workers have performed a complicated dance that happens around the world day after day, year after year.

We can also learn something from how technological advancements that maximize safety and profit stimulated cruise line evolution and growth. For example, safety and economy were improved when:

- Early ships went from burning coal to burning oil.
- Radios and, decades later, radar and Global Positioning Satellite (GPS) systems, were first introduced.
- One of the Coast Guard's main missions became life safety and monitoring icebergs in the Atlantic.
- Ship construction went from using rivets to steel welding.
- Seasickness was largely overcome by building larger ships with active stabilizers and with the use of drugs such as Dramamine.
- Fully functional mini-hospitals and, later, landing zones for emergency helicopters were added to ships.

A new kind of cruise ship was launched just a few years ago. The *World of ResidenSea* is the first cruise ship to offer 120 private luxury homes where passenger residents can live year-round. The company's motto: "See the world without ever having to leave home." Homes range in size from 1,500 to 3,000 square feet and offer outside terraces. Some offer three levels, each custom designed. The owners provide their own furniture, art, books, and clothing.

The $375 million ship is the first of two sister ships and the first major innovation in cruising in well over 50 years. Her 60,000 tons also provide 80 guest staterooms for families and friends and extensive support services to the residents. Features include a mini-shopping mall, swimming pools, small private restaurants, and a full-service gymnasium and business center. There is also an extensive hospital and health club, and crew quarters with their own recreation areas.

The future of the cruise line industry looks very good. Some of the larger cruise lines have the size and asset base to be key players in financing orbital space tourism, along with their proven ability to profitably and safely operate ships in extreme environments, while providing luxurious and memorable experiences for their passengers. Someday there will be half a dozen large orbital ports servicing 100 orbital and lunar cruise ships; hundreds of orbital super yachts and sports facilities; plus all of the industry, science, and military facilities, and ships. Studying seaports and the cruise lines will provide a vital frame of reference to design an on-orbit zoning plan.

Clearly the cruise lines will become the space lines of the future. In the meantime, the super yacht industry will take the lead in developing the first stage of real space tourism.

The Orbital Super Yacht Breakthrough

The first privately financed and operated passenger ship in Earth orbit will be an elegant orbital super yacht. She will be owned by one of the world's largest corporations or wealthiest individuals and will become the new symbol of power and prestige.

The history of super yachts can be traced to the Egyptian pharaohs whose royal barges slowly cruised the Nile. These beautiful floating palaces provided luxurious travel for the royal family, and also served as symbols of their supreme power. For 5,000 years, large yachts have served as symbols of wealth and power.

In today's terms, super yachts are also luxurious, more than 150 feet in length, the average cost is well more than $40 million to construct. They typically carry 10 to 20 guests plus crew. The largest super yachts are more than 300 feet long, cost more than $100 million to build, and accommodate 20 to 30 guests in luxurious staterooms and elegant public areas adorned with fine art and furnishings costing more millions of dollars. Some of the larger super yachts have landing pads for their own helicopters. It costs millions of dollars per year to crew and maintain them at exclusive yacht clubs. As world wealth grows, the 70 private yacht-building companies in Europe and the United States cannot keep up with the demand for larger and more elegant super yachts.

I was surprised at how many hundreds of these beautiful super yachts there are worldwide. I was also surprised at the high level of advanced technology used to build, outfit, and operate them. Award-winning yachts balance sophisticated design with art and advanced technology. New construction materials, advanced electronics, jet water engines, hull designs, safety features, and space-saving interior design, contribute to an operational efficiency that is amazing. Their engine rooms and bridges resemble the bridges of futuristic star ships. The interior decor and furnishings are as stylish as anything on land.

The best magazine on super yachting is called *Boat International USA* (*www.boatinternational.com*) and there is an excellent design-oriented magazine called *Yacht Design* — both available in the sailing section of major magazine stands. If you look through these beautiful magazines, picture these surroundings applied to a new type of space-cruising craft. It can be done!

Beauty aside, the most important aspect of super yachts is that they do not exist to make a profit. No one buys a ticket to take a trip on a yacht. They exist as status symbols for the wealthy and powerful, and for rewards, marketing, and political purposes. Yachts are an entirely different financing and operational model from cruise ships.

Realizing this unique characteristic of super yachts led to an extraordinary "ah-ha" moment. I realized that a similar incentive model would drive the development of the first phase of orbital vehicles. Substituting the ancient and powerful desires for prestige, power, and status for the more mundane desire to make money in order to cover costs, can win the way to space. With orbital super yachts, entry to orbit becomes a new game many powerful people will want to play.

I will always remember the day I had the orbital super yacht conceptual breakthrough: November 24, 1996. It was one of those rare "eureka!" moments when you realize you've just made an important discovery that will forever change your life and possibly make a positive difference for humanity. As an explorer on the design frontier, I have been fortunate to have a few other conceptual breakthroughs and design discoveries, but none to date as significant as the orbital space yacht. It is a powerful and logical course to pursue in successfully building the space tourism industry.

Since that discovery, I have studied the yachts and their culture, attended incredible yacht shows, met with many people who love yachting, and discussed the concept of orbital yachting, finding them a supportive audience. I have now devoted years to creating the first space architectural design of an orbital super yacht I call *Destiny*. The design has become a treasure trove of questions and research, and I am working to bring *Destiny* to life by the year 2030. Her design is premiered later in this book.

Where Does One "Tie up" in Orbit?

First we find the Earth equivalent for this function: yacht clubs. What can they tell us in our search for a model to develop a fundable orbital facility? Earth-based yacht clubs are entirely funded and supported by membership fees, with some revenues generated in their restaurants, bars, shops, and other member services. They are internationally located, with more than 300 major yacht clubs all over the world, especially in Western Europe, Florida, and California. The clubs provide moorings for as few as 20 yachts to more than 200. They also provide important services such as fuel, food, crew, and basic maintenance as well as parking for owner and passenger cars. Dry dock maintenance and yacht construction are done at separate private shipyards.

If six to eight mega-corporations and/or individuals commit to orbital yachting as the next great prestige trend, they may be persuaded to form the first orbital super yacht club and finance its assembly and operation. This club will provide a place in space and future location for the first privately owned orbital port.

There is also a defined culture involved with yachting. It is an affluent, sophisticated, educated, and technology-oriented culture that wants to be special and, in many ways, secluded from the rest of society. These people have made it, and they want to show off to each other. Yacht races are the perfect example of this. Wealthy people, companies, and sponsors race each other not for any practical reason, but to be challenged, to have fun, and to prove who is the best of the best. Along the way, new technologies and capabilities are also developed that have many spin-off benefits.

According to *Forbes* magazine, in 2003 there were close to 400 billionaires worldwide. The majority of them will own super yachts or use them for marketing and political purposes. Once you own half a dozen luxury homes, and your own private jet, the next acquisition of the super rich is a luxury yacht.

Some of the extremely wealthy and powerful, when convinced that orbital yachting is the next very cool thing to do and they become inspired and engaged, happen to own major media outlets that could provide needed media attention. Ted Turner, founder of CNN, donated $1 billion to the United Nations and supported Jacques-Yves Cousteau's exploration and popularization of Earth's oceans. Turner was a great yachtsman, having won the 1977 America's Cup in his yacht, *Courageous*. Another is Larry Ellison, founder and CEO of Oracle. Roy Disney Jr., whose father and uncle Walt Disney founded the Disney Corporation, is an avid yachtsman. There are dozens more who are directly involved in advanced technology, entertainment, and aerospace plus hundreds more from all other world industries. Whom do you know?

Still another reason to focus on the yachting community is the allure of personal and environmental enlightenment that would occur through an orbital experience. Author Frank White's 1986 book *The Overview Effect*, describes the orbital experience as providing a growing awareness of our position in the cosmos and a deepening of respect for nature and other people. With the powerful yachting culture, this effect could have profound social benefits.

Four Phases of Space Tourism Development

In 1990 two Englishmen, David Ashford and Dr. Patrick Collins, published the first book on space tourism. It was titled *Your Spaceflight Manual: How You Could Be a Tourist in Space within 20 Years*, published by Headline Book Publishing in England. This pioneering work did an excellent job of introducing the concept of space tourism and of outlining their plan for developing the industry. Their four phases of industry development were modeled from the development experienced by the airline industry from 1923 through the mid-1970s.

With their permission, I have adapted their ideas to include my vision of the evolution of space tourism based on the super yacht and cruise line models. I hope I am conservative with my estimates of when phases will start and end.

The Pioneering Phase: 2001 to 2020

I consider Dennis Tito's flight in April 2001 and the huge media coverage of his off-world adventure to be the beginning of the Pioneering Phase. During this phase, wealthy individuals will buy tickets or corporations will sponsor celebrities to fly with the Russians to the Space Station. Lotteries, promotional contests, and reality television shows like *Survivor* will send tourists to space, too. The first privately funded suborbital flight will win the Ansari X Prize competition with a possible second X Prize that requires orbital flights to follow.

Space tourism themed expos, conferences, award shows, theme parks, resorts, orbital and space cruise immersive simulation centers, movies, television shows, books, and more sophisticated websites and virtual worlds will expand the public's awareness of, and

Four Phases of Industry Development

2001	2010	2020	2030	2040

Pioneering Phase

Wealthy private citizens and lottery winners

ISS conversion into first private orbital yacht

Exclusive Phase

Second generation of orbital yacht

Begin orbital sports activities

The Space Guard Service begins operations on orbit

Orbital Yacht Racing

Mature Phase

Third generation of orbital yachts starting with *Destiny*

Development of first orbital yacht club — *Eagle*

Mass Market Phase

Space cruise ships and lunar cruising

First sports stations and orbital ports

The evolution of the Earth-based immersive space/lunar/Mars simulation industry

participation in, the space tourism industry.

Advances will be made in technical research and testing of reusable orbital access vehicles (OAV). Progress will be made in the important areas of international law, regulations, and the development of ground-based infrastructure, launch complexes, and Earthports.

By 2020, nearly 1,000 private citizens will have traveled off world. The novelty will have worn off. Millions of people will have taken simulated space vacations and adventures on Earth.

The Exclusive Phase: 2020 to 2030

Beginning in 2020, some countries, corporations, and wealthy individuals and foundations will form orbital yachting clubs. The members will pay for the design, transportation to orbit, assembly, and operation of the clubs. Members will do the same for their private orbital super yachts. Like their ocean-based forebears, the orbital clubs and yachts will cater to the ultra-rich; royalty; politicians; film; music; and online stars, sports heroes, and other celebrities. They will be invited to enjoy orbital yachting and eventually orbital yacht races with the worldwide media capturing their adventures. The first superstar wedding and honeymoon in orbit will cause a huge media sensation.

Some people not on the extreme "A" list will pay several millions of dollars to have the privilege of going to experience the glamour and unique aspects of orbital tourism. This phase will see space tourists continue on to the Moon and beyond, once the thrill of LEO tourism is over.

Sports in space could significantly increase the number of people going in this phase. This would require much more orbital infrastructure that will facilitate the financing of orbital ports, resorts, sports facilities, and other facilities and services needed for a maturing market.

The Mature Phase: 2030 to 2040

By this time, far greater numbers of the wealthy will be going off world — up to 10,000 each year. The orbital cruise lines will offer their first cruises to large groups of tourists and support people. The early orbital cruise ships will carry up to 100 passengers and crew members but will grow to carry more than 500. Busy orbital ports with resorts, shopping centers, hospitals, and recreation areas will be necessary, as well as orbital farms, assembly areas, fuel areas, storage, security, and all the other activities and facilities typically found around ports. Lunar cruising will be established with growing numbers of adventure vacationers and tourists landing on the Moon to stay at small resorts and visit historic lunar sites. Orbital and lunar sports will be established. Orbital super yacht races modeled after the America's Cup will become annual events.

The Mass Market Phase: 2040 and Beyond

As a frame of reference, remember that today's cruise industry services 12 million passengers each year on overnight cruises. Our definition of "mass market" is based on this number, rather than the far greater numbers of passengers taking airline flights.

As the space tourism industry passes through its first three phases, Earth-based market demand, infrastructure, and safe access to orbit will also be increasing and improving. The orbital ports, sports stadiums, and casinos will have hotels costing less than the space cruise ships and lunar resorts. Their promotional and group-rate programs will create a mass market drawn to orbit.

An orbital tax system will be established that motivates governments to support expanding the infrastructure for off-world travel. The financial community will be competing for profits from space tourism. Private enterprise will build the Las Vegas of orbit. Like the Earth-based Vegas, it will grow into a small orbital town and then into a city, creating the first off-world, self-sustaining space colony of at least 10,000 people and visitors.

By the end of this final phase, one million people each year will travel off world. The year 2075 could see 3,000 to 5,000 tourist and sports fans going every day — 1.1 million to 1.8 million space travelers per year.

Conclusion

Space tourism is all about the experience. It is not about launches or science or exploration. If you remember anything from this book, remember that.

By taking my cue from the successful and experience-centric cruise line industry, I have worked to shape a vision that addresses time, markets, players, and the desires that drive humans to want to do something new.

I like to imagine that someone someday will write a book that promotes the "wild idea" of star tourism and proposes a new industry in which humans tour the stars by going beyond the solar system. Perhaps this writer will build on the same four phases I have used here and make their quest an easier one.

In this chapter I touched on the market for space tourism and mentioned potential consumer markets including women and the adventurous wealthy. In the next chapter, I explore these market factors more closely, including space-related adventures here on Earth. Do they count as space tourism? Sure! Read on.

CHAPTER 2
IS THERE A MARKET?

Introduction

Is there a market for space tourism? This is one of those questions that is both easy and difficult to answer. Having personally witnessed the enthusiasm of thousands of people wanting to go into space, including several who are very wealthy, I know there is a market.

One of the first steps in identifying a market is to have a clear definition of the product you are marketing. In our case, we define the experience. The Space Tourism Society has developed a definition of the space tourism product:

- Earth-based simulations, tours, and entertainment experiences
- Cyber- and tele-tourism, such as driving a rover on the Moon from Earth
- Earth orbit experiences
- Beyond Earth orbit (such as lunar and Mars) experiences.

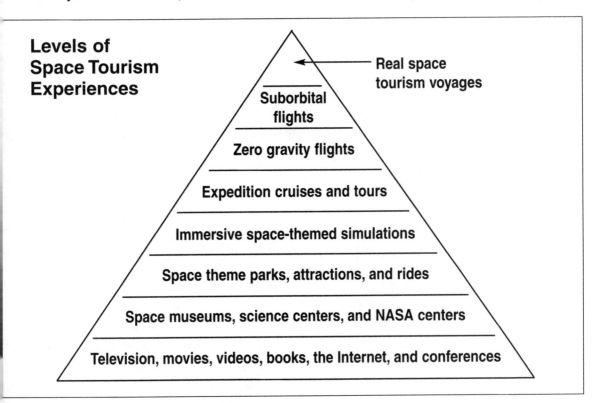

Levels of Space Tourism Experiences

Real space tourism voyages

- Suborbital flights
- Zero gravity flights
- Expedition cruises and tours
- Immersive space-themed simulations
- Space theme parks, attractions, and rides
- Space museums, science centers, and NASA centers
- Television, movies, videos, books, the Internet, and conferences

According to this definition, there are already millions of people each year who have had space tourism experiences by visiting space museums or NASA visitor centers; witnessing Space Shuttle launches and landings; spending a week at a space camp; flying in a zero gravity aircraft; going on astronomy-themed cruises; or participating in space conferences, expos, foundations, societies, and businesses.

Ask yourself if you think there is a market for space tourism. Do you know someone who would like to go if it were safe and affordable? Would you like to go? According to a half dozen professional market surveys done in Japan, the United States, and Europe between 1996 and 2002, millions of people answered "yes" to the question, "Would you like to take a real space vacation?"

We already know of three wealthy men who paid millions to the Russians to fly to the International Space Station and spend a week aboard the station, having the times of their lives. There are at least a half dozen other wealthy individuals who have the resources and who have expressed strong interest in going once the Space Shuttle fleet is back in operation.

The Earth-based space tourism market and community will experience significant growth by the year 2010 as the next generation of immersive space cruise, space race, and adventure simulation facilities open. At the same time, a growing number of private citizens will have either flown to, or are in line to fly to, the International Space Station.

A Long History of Interest in Space Travel

There is a long history of private citizens interested in space exploration and travel. Writers in the late 1800s used their imaginations as the vehicles of their day to take themselves and their readers to new worlds beyond Earth. Many were fascinated by real world science, new discoveries, new theories, and new inventions. They enlisted the scientists and inventors of their day to critique their writing, bringing science to fiction, and in doing so they invented the science fiction genre.

The great French writer, Jules Verne, wrote and published his book *From the Earth to the Moon* in 1865 while the American Civil War was just ending. One of his many visionary predictions suggested that the American Florida peninsula was an excellent location from which to begin a voyage to the Moon. In 1969, *Apollo 11* lifted off from Florida for its historic landing on the Moon, less than 60 miles from the point predicted by Verne 104 years previously.

In 1902, a French movie pioneer, Georges Melies, created *Le Voyage dans La Lune* (*Voyage to the Moon*), one of the world's first feature-length films. The movie created a sensation in France and drew worldwide attention. Agents of famous American inventor Thomas Edison stole copies of the movie and showed it in America. In fact, Edison made the first fortune in the movie business with Melies' space-themed adventure, making it the *Star Wars* of its day and inspiring a string of other space-themed movies.

In the 1930s, a generation was captivated by radio and television serials of *Flash Gordon* and *Buck Rogers*. These heroes with their spaceships and flying backpacks explored space and fought aliens determined to conquer Earth.

In 1950 the movie *Destination Moon* was released, depicting the first realistic human mission to the Moon. It was produced by George Pal, based on a story by famed science fiction writer Robert Heinlein.

In 1955, Walt Disney premiered his science factual programs including the three-part television series *Man in Space, Man and the Moon, Mars and Beyond*. He hired prominent rocket scientists Wernher von Braun and Willy Ley, along with space artist/astronomer Chesley Bonestell, to design realistic vehicles and missions to the Moon and Mars. These television shows were big hits with the public.

Many of the *Apollo* engineers and astronauts credit these programs and the space adventure books of their day for first introducing them to the concept of space travel and inspiring their interests.

In 1967, Baron Hilton, then president of the Hilton Hotel Corporation and one of the most experienced and respected men in the hotel industry, presented a well-conceived and researched paper to the American Astronautical Society (AAS) titled "Hotels in Space." The paper outlined the issues surrounding the potentials for building a hotel in Earth orbit and a 'Lunar Hilton." This was a serious perspective from an industry leader who had a strong reputation for knowing market trends and interests.

In April 1968, the movie *2001: A Space Odyssey* premiered, stunning audiences around the world with its incredible visual effects and designs for advanced space stations, Moon bases, and exploration vehicles. Both Hilton and Pan Am Airlines paid the movie producers to have their company names on screen. (The Hilton name appears on the huge rotating wheeled spaceport/hotel, and Pan Am's appears on the sleek Space Clipper ship.)

In 1983, author Tom Wolfe's best-selling book, *The Right Stuff*, reignited interest in the space program. It inspired the movie by the same name and complemented the American Space Shuttle fleet's early days of service.

Since the formation of NASA in 1958 and President Kennedy's famous 1961 speech about sending a man to the Moon and bringing him back safely by the end of the decade, NASA's *Mercury, Gemini*, and *Apollo* programs stirred the frontier spirit and imaginations of the American public. Legendary journalist and television news anchor Walter Cronkite became the unofficial spokesman for the space program. His credibility and enthusiasm for the space race to the Moon drew the attention and support of viewers, while at the same time informed and taught the public about space travel.

The *Apollo* program stimulated the interest of many of today's aerospace engineers, astronauts, program managers, and business executives who are running the Space Shuttle program and building and operating the International Space Station. My own interest in space was stimulated by the *Apollo* landings and the glory days of the space program.

Surveying the Market

Any business must understand its market if it is to prosper. One way to do this is to conduct professional market surveys of potential customers. For the space tourism industry, some significant pioneering research has already been done.

Patrick Collins, Ph.D., is the leading pioneer in designing and conducting market surveys focused on space tourism. The surveys he conducted in Japan indicated a highly positive response to the concept of space tourism. The thousands of people interviewed were also willing to spend more on a space vacation then a normal vacation. All of his studies can be found on the SpaceFuture.com website.

In April 1997, the marketing firm of Yesawich, Pepperdine, and Brown (YP&B) based in Orlando, Florida, conducted a telephone survey of 1,500 homes nationwide, asking questions pertaining to space tourism. Forty percent of those surveyed said "they would look forward to an out of this world vacation experience." That represents almost 80 million Americans.

"Space travel represents one of the next great horizons in leisure travel as more and more consumers seek extraordinary travel experiences," said Peter Yesawich, president and CEO of YP&B and coauthor of the survey. "It's the logical sequel to trekking in the Himalayas or camping with headhunters in Borneo."

Space Adventures conducted its own poll on space travel in 2000, with 2,022 participants across the United States and Canada. Although the survey focused primarily on suborbital flights, it found that 86 percent of those polled were interested in space travel for tourism and leisure, with 10 percent earning enough to go through with it given the chance. There were very clear projections that thousands of people — between 5,000 and 10,000 per year — would participate in suborbital spaceflights priced at $100,000 a flight. Their findings showed it's a billion dollar market.

During the week that Dennis Tito was in orbit, there were several surveys conducted to ascertain the public's opinion of his flight. On April 30, 2001, the cable news station MSNBC asked the question, "What do you think about putting paying passengers in space?" They received 55,005 responses with the following breakdown:

66%   "I would take the trip, if I could afford it."
19%   "I don't think the time is right for space tourism"
14%   "I wouldn't go, but I see nothing wrong with others going"

Following Tito's return, *Aviation Week & Space Technology* magazine conducted a poll asking, "Did Dennis Tito's seven-day International Space Station excursion prove good for the future of space exploration?":

79%   "Yes, it stimulated public interest, which is crucial for support of such an expensive program."
21%   "No, tourists divert the crew from their responsibilities and compromise flight safety."

While these market surveys have made an important contribution to our understanding of the general public's interest and support of real space tourism, additional surveys must be designed and conducted that focus on the wealthiest segment of American society. Fortunately, one has been completed, and it was paid for by NASA.

In 2002 an important market study and industry analysis was commissioned by NASA to forecast America's need for access to orbit in the year 2020. It is called *The ASCENT Study: Understanding the Market Environment for the Follow-on to the Space*

Shuttle. The study was conducted by the Futron Corporation, a Maryland-based aerospace consulting group (*www.futron.com*). The director of the study, Derek Weber, a long-time supporter of space tourism and the first lifetime member of the Space Tourism Society, provided the following information about one section of the study, which was focused on space tourism. The section is extracted from an interview for Space.Com:

"Poll: America's Wealthy Willing to Pay Top Dollar for Spaceflight"
(*www.space.com/news/space_poll_020520.html*)
By Tariq Malik, staff writer, May 20, 2002

"The survey, which polled hundreds of affluent U.S. residents on their interest in space travel, is the latest in a string of market research studies looking at the public's interest in space tourism. What sets it apart, pollsters say, is its focus on individuals with the financial means, and not just the desire, to make the trip."

Zogby International conducted telephone interviews of 450 American adults whose yearly incomes exceed $250,000 and net worth is approximately $1 million or more. All calls were made from Zogby International headquarters in Utica, New York, from January 6 through January 27, 2002. The margin of sampling error was ± 4.7 percent. Survey participants were confined to those who could at least potentially afford the high prices of this leisure activity (which is expected to cost around $100,000 for the lowest-cost package).

The poll was designed to measure the level of interest in public space travel; the willingness to pay for specific space travel options; and an array of other relevant information concerning lifestyle choices, spending patterns, and attitudes toward risk. Some of the more interesting findings included:

- By 2021, commercial space travel could amount to an industry worth more than $1 billion.
- Suborbital space travel is a promising market with space tourists being rocketed 50 miles into space, at an assumed cost of $100,000, and experiencing much the same kind of 15-minute experience of exhilaration, weightlessness, and seeing the Earth below as did Alan Shepard, America's first astronaut.
- Up to 19 percent of those interviewed indicated that they would be likely to take part in such an experience when it becomes available to the public, assuming they could meet the medical and other requirements. Futron's forecast for suborbital space travel projects that by 2021, more than 15,000 passengers could be flying annually, representing revenues in excess of $700 million.
- Orbital space travel is also a promising market — Futron's forecast for orbital space travel projects that "by 2021, 60 passengers may be flying annually, representing revenues in excess of $300 million."
- In the case of two-week orbital flights to an orbiting space station, a surprising seven percent of those wealthy individuals polled said they would be willing to pay today's price tag of $20 million. The figure approaches 16 percent if prices come down to a "mere" $5 million a ride.
- Of those surveyed, 52 percent indicated that post flight physical discomfort (e.g., dizziness and difficulty standing) would make no difference in their decision to purchase a two-week orbital flight.

- The most important thing about on-orbit destinations is options. Futron estimated that an increase in demand would result from having accommodation options at both the International Space Station and a commercial on-orbit facility available, yielding a total of 553 passengers over the forecast period — a 32 percent increase over the forecast with the International Space Station as the sole on-orbit destination option.

Derek Weber reported that astronaut Buzz Aldrin was able to use some of the findings to help him make his case as a Commissioner of the President's Commission on the Future of the U.S. Aerospace Industry. As a result, for the first time ever, a blue ribbon government report includes public space travel as part of its recommendation! The final report of the commission points out that the aerospace business needs the space tourism sector to take off.

"Where Do I Sign Up?"

Whenever an invitation or program is announced offering a way to get into space, thousands respond. In 1951, the Hayden Planetarium in New York City invited readers of their magazine to fill out a form registering them for a trip into space. Over the years they received more than 400,000 responses.

Since the late 1960s, every time that NASA announces there are 15 to 20 open astronaut candidate positions there are at least 25,000 applications submitted, of which thousands come from individuals with Ph.D.s and from senior military officers.

In 1969, Pan Am announced it was taking reservations for a trip to the Moon. It was a publicity program called the First Moon Flights Club, which received more than 90,000 responses. Pan Am was both surprised and dismayed by the large response.

Thirteen years later, in 1982, the head of NASA began the Spaceflight Participation Program to fly private citizens aboard the Space Shuttle. As noted in an earlier chapter, NASA received more than 11,000 applications.

Also in 1982, Bob Citron, who later founded the successful Spacehab company and cofounded Kistler Aerospace, created an exciting program called the Space Shuttle Passenger Tour. Working with Society Expeditions, a well-respected adventure vacation tour operator, their plan was to lease one of the Space Shuttles to take paying passengers into orbit on a number of dedicated missions in a specialized passenger module. This was the first credible and serious effort to develop real orbital tourism by an experienced adventure tour operator. They received thousands of inquiries and, by 1984, had just over 200 flight deposits for $5,000 each. However, after financing fell short and NASA refused to open the Space Shuttle to tourists, the deposits were returned.

That same year, I created a concept and design called "Space Resort," which proposed building a full-scale mockup of a 200-cabin orbiting space resort that would provide passengers with two- to five-day immersive simulated cruises on Earth. I modeled the space resort cruise experience after a Princess Cruise Line cruise ship. After a few hours of training and after changing into comfortable flight clothing, passengers would ride an open elevator up a full-scale gantry tower and cross over into the cargo bay of a full-scale Space Shuttle standing on a launch pad. After a thrilling liftoff and docking with the wheel shaped rotating Space Resort, *Athena*, the passengers would then ride hidden elevators in the Solid Rocket Boosters (SRBs) down to the Resort Hub beneath the launch pad to begin their cruise. After three to five days immersed in the cruise simulation, the voyagers would enjoy an elaborate return-to-Earth simulation, followed by several hours at "Cruise Control Center" to decompress from their off-world vacation.

The reason I note the Space Resort project here is because of the large number of people who signed up to participate in the funding and development of the project. The company I formed to build the resort formed a joint venture with Mitsubishi Corporation and built the full-scale vertical Space Shuttle stack at a space theme park we helped plan called "Space World." The park opened in Japan in 1991 and welcomes one million visitors every year.

Back in 1982, the concept of space tourism was entirely new and revolutionary. The Space Resort project drew the attention of the media, as well as support from NASA and the aerospace industry. We formed an advisory board of famous and influential people from the entertainment, resort, and finance communities. Members included the retired head of NASA during the *Apollo* landings, Tom Paine, and Dr. Buzz Aldrin. Dr. Aldrin has been a friend and business associate to this day. Our advisory board also included the retired chairman of the board of the Disney Corporation, Don Tatum; the creator of *Star Trek*, Gene Roddenberry; and Charles Weber, the former president of Lucasfilm, Ltd., whose founder created the *Star Wars* films and cocreated the *Indiana Jones* films. We also received strong encouragement from Dr. Carl Sagan, the world-famous astronomer and creator of the television miniseries, *Cosmos*.

An unexpected reward during the Space Resort development experience happened in 1984 at a meeting/presentation with Steven Spielberg's associate, Frank Marshall, at their office on the Universal Studios lot. While walking down a hallway to Marshall's office I literally bumped into Mr. Spielberg. I was carrying a box containing the Space Resort model and a large portfolio with display boards. He asked me where I was going and what was in the box. After a brief explanation, Spielberg replied, "Wow, I want to see that!" and being a gentleman, helped me carried my model box into Marshall's office and then sat down to watch most of the presentation.

Each of the accomplished people I just noted, told me they wanted to go into space or back to space. Each said they wanted to take their families or grandchildren to a real space resort someday and each in their own way helped with that project. I now have a new and more sophisticated version of the Space Resort project back on the drawing boards.

In 1990, a British firm announced a contest for a British citizen to fly into space. The contest attracted 13,000 participants, with the winner flying to *Mir* in late 1991 for a week-long visit.

In the mid-1990s, an American firm used the Internet to announce a nationwide lottery for a vacation flight to *Mir*. Within the first week they claim to have received more than one million inquiries on their site, though legal issues prevented the lottery from taking place.

In August 2000, Mark Burnett, the creator and executive producer of the mega-hit television series *Survivor*, announced that he was planning a *Survivor* show called *Destination Space*, in which the winner receives not only $1 million, but also a trip to the *Mir* Space Station. NBC committed $40 million to that incredible show. Unfortunately, the

Mir was de-orbited and plans for the show were shelved.

As of early 2004, one company in Japan, two in Europe, and three in the United States are exploring the legal and marketing issues of conducting a nationwide lottery with the winners flying to the Russian part of the International Space Station.

These examples of thousands of people who have signed up for a chance to go into space demonstrate a small but consistent and real market. Should the producers of some of the proposed lotteries succeed with millions of people buying tickets, we will have another indication of global interest in space tourism.

The Earthside Market

Remember that our definition of space tourism includes "Earth-based simulations, tours, and entertainment experiences." Since the mid 1960s, hundreds of millions of people have had space tourism experiences around the world. They visit space museums, NASA visitor centers, space camps, and a variety of other space-themed attractions and locations. About 14 million people annually visit the two Universal Studios movie theme parks in California and Florida. However, more people visit the following space-themed facilities each year:

The National Air and Space Museum	10,000,000
Kennedy Space Center	3,000,000
Johnson Space Center	1,000,000
Space Camp in the U.S.	300,000
Total Direct Market	14,300,000

According to NASA, well more than 25 million people have personally witnessed the launch of a rocket or Space Shuttle from the Kennedy Space Center since the mid 1960s. Some people have witnessed dozens of launches and landings of the Space Shuttle.

John Glenn's flight in October/November of 1998 was a great boost to the image of NASA and the overall American space program. He proved that people in their seventies could fly into Earth orbit. He attracted huge international media attention and generated strong good will. He and his crew received the largest ticker tape parade in New York since U.S. troops returned from *Desert Storm*.

In 1997, the successful Mars *Pathfinder* mission received worldwide attention. Its NASA website received more than 115 million hits in one day — a world record at that time.

Investments in space museums and NASA visitors centers have had significant increases since the beginning of the 1990s. The Houston-based Johnson Space Center's (JSC) $75 million Visitor Center opened in 1993 and, in 1996, the $45 million Apollo/Saturn V Museum opened at the Kennedy Space Center. The JSC facility is now undergoing a $100 million expansion.

"Almost 300 science and technology centers in the United States welcome 115 million visitors a year — a threefold increase in the past decade alone," said Bonnie Van Dorn of the Association of Science-Technology Centers (ASTC) in *Time* magazine on November 15, 1999. Almost all of these centers have space-themed areas and exhibits.

Many have large screen IMAX theaters that show space films such as *The Dream Is Alive*, *The Blue Planet*, and *Space Station*, which was narrated by actor Tom Cruise. All were filmed in part by the astronauts on Space Shuttle and International Space Station missions.

The astronomy market is also growing. Millions of people around the world enjoy astronomy as a hobby. They buy telescopes, cameras, books, and magazines and take astronomy classes, themed cruises, and vacations. There are nearly 1,000 observatories and planetariums in the United States. Several million people per year take tours and participate in stargazing. Most planetariums have space and astronomy shows. One of the newest, the renovated Hayden Planetarium in the Rose Center for Earth and Space at the American Museum of Natural History in New York City, features a sphere 95-feet in diameter and a breathtaking astronomy show. These ground-based space tourists spend a significant amount of time and money on their hobby.

Space-themed simulation rides offer millions of people each year a glimpse of what real space exploration and tourism could be. In the 1960s, many of our parents and those of us who were kids at the time, rode the exciting Disney *Trip to the Moon* simulator in Tomorrowland. The ride was renamed *Trip to Mars* in the 1980s. Since the mid-1980s *Star Tours* at all Disney parks have taken more than 100 million people on exciting space tourist rides through the *Star Wars* universe. The *Space Shuttle Experience* at Six Flags in Chicago offers a full-scale mockup with a theater in the cargo bay. As of the end of 2003, this attraction has taken more than 10 million people on an exciting ride to and from the Moon.

In October of 2003, Disney opened *Mission: Space* in Epcot Center at Walt Disney World in Florida. This $120 million pavilion was sponsored by Hewlett-Packard. It takes visitors on a simulated ride to Mars and returns them to Earth. NASA and retired astronauts were consultants on the design of the attraction. Disney expects between three and five million people to take the *Mission: Space* experience each year and, due to its popularity, is considering replicating it at other Disney parks around the world.

The Adventure Vacation Market

In the early days of the space tourism movement, we always used examples of adventure vacations to create empathy and credibility for our ideas. This was usually successful and is even more so today with private citizens paying to take deep-sea submersible dives to the *Titanic*, flying in Russian Mig fighter jets, taking cruises to Antarctica, sailing in Russian nuclear-powered ice breakers to the North Pole, and taking passenger submarine trips. The true adventurer can parachute to and raft down the Amazon river, climb Mt. Everest, hunt for brown bear with only a bow and arrow, explore active volcanoes, and have many other extreme adventures. In fact, by comparison, space tourism seems tame.

People who are interested in taking trips to Antarctica are the kind of people who will be interested in space tourism. Adventure travelers are very demanding not only in wanting luxury, but also in expecting to learn from world-class experts. They tend to be highly educated and love discovering new places, people, costumes, food, music, and sights, and they really want to know the details and what the experts think will be coming next.

Today, people also seek new, exciting, and meaningful vacation experiences that may not need to provide as much of an adrenaline rush as those just mentioned. These mature affluent, well-educated, and well-traveled people have more money than time, so they seek vacation experiences that are memorable and personally rewarding. Expedition tour groups take hundreds of thousands of people on small special expedition cruise ships to unique destinations including Antarctica, Easter Island, and the Galapagos Islands. The tours feature safe, luxurious accommodations with world-class experts along as guides.

The exotic and expedition tourism market has been growing at an accelerating rate since 1990, coming close to $1 billion in revenue in 2001. Jerry Mallett, founder and president of the Adventure Travel Society, sees the industry having almost unlimited growth potential, with space tourism becoming the new adventure. As the key speaker at the Space Transportation Association conference on space tourism in Washington, D.C., in June 1999, he noted that there were at least 10,000 adventure travelers who would spend at least $100,000 each for a suborbital trip, and many would spend much more for an orbital trip.

Space Adventures has already received more than 200 deposits at $5,000 each for suborbital flights that will ultimately cost $98,000. This represents almost $20 million in space tourism business. A suborbital flight that duplicated *Mercury* astronaut Alan Shepard's historic flight would not go into Earth orbit and would last only about half an hour, but it would provide nearly 15 minutes of zero gravity, and you would clearly see a dark sky above and the beautiful curvature of the Earth below. The entire experience would last about a week, starting with training and briefings, the flight, and then a few days of postflight debriefings and relaxation. All who go to the edge of space through this exciting program will earn their astronaut wings.

In a similar vein, 1997 marked an historic year for the passenger submarine industry. More than one million passengers took a one- to two-hour dive under water in one of about 50 privately-owned tourist submarines worldwide. In the early 1970s, the pioneers of private submarines and the underwater tourism industry faced many of the same challenges we face today in developing space tourism. They succeeded, and are now planning larger submarines and underwater resorts with dozens of luxury guest rooms. Now, several super yachts are specially designed and built to have the capability to completely submerge and cruise at 1,000 feet below the surface.

Another important connection is being forged between the space tourism community and the eco-tourism community. Why? Because space tourism will be the ultimate eco-tour. Orbiting the Earth every 90 minutes will allow tourists to see the majority of the planet on a seven-day space cruise. I have talked with astronauts who say the view of Earth from orbit is the most magnificent experience they have ever had. Many have said that, once they saw Earth from orbit, they felt far more aware and concerned about its fragile ecosystem. In fact, the ecology movement was fueled by the incredible photos of the blue and white Earth against the deep black of space taken by the *Apollo* astronauts on missions to the Moon.

The Return of a Market

During the 1930s a new form of luxury travel was maturing. In the skies flew huge, 800-foot-long airships called the *Graf Zeppelin* or the *Hindenburg*. They crossed the Atlantic a

Dennis Tito, the world's first Space Tourist, boarding the International Space Station on April 30, 2001. His first words as he floated through the hatch: "I love space."

Television image courtesy NASA

The Space Tourism Society's Space Tourism Pioneer Awards held on April 27, 2002, in Santa Monica, California. "Orbit" Award honorees on stage during the show finale.

Dennis Tito receiving the first "Dennis Tito" award during the show.

Photos by Larry Evans
courtesy of the Space Tourism Society

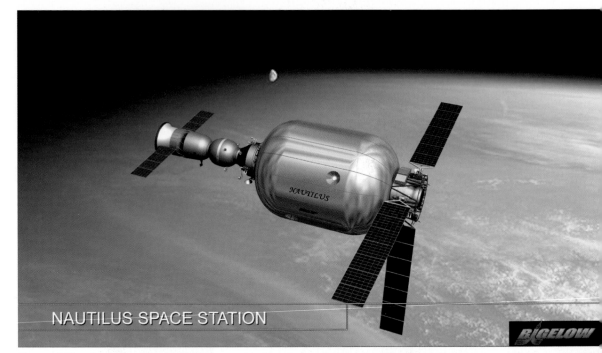

NAUTILUS SPACE STATION

One of the front-runners in the space tourism stakes is Bigelow Aerospace of Nevada. The company opened its doors in April 1999 with the long-range vision of developing an aerospace business that would participate in commercial space flight. The core business infrastructure is provided by the existing businesses owned by Mr. Bigelow, including Budget Suites of America. Pictured above and below are the designs currently being investigated at Bigelow's Nevada facility.

NAUTILUS SPACE HOTEL

Photos courtesy Bigelow Aerospace

Bringing the adventure of Space to the people of Earth

The 1997 cover of the Space Adventures brochure. It is their first brochure and announces opportunities for space tourism. The view shows the Pacific Ocean with clouds and volcanic smoke (bottom center) as taken aboard the Space Shuttle *Discovery*.

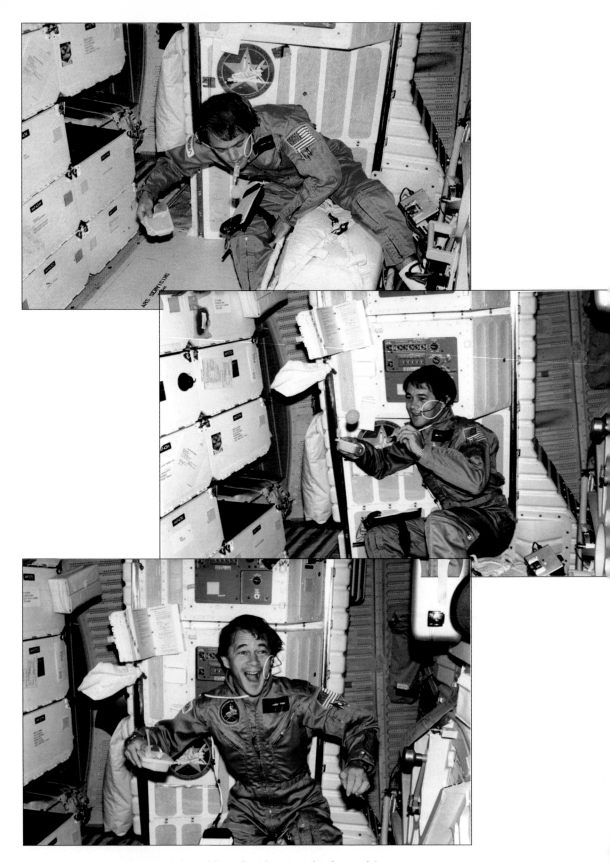

Astronaut Joe Allen is shown playing with orange juice that has formed into spheres in 0-g. Photos taken aboard the Space Shuttle *Columbia*.

A floating and zero gravity sports sphere inside a futuristic space resort. Research and design work from 1993 to 1996 showed that this sort of facility was commercially viable.

Artwork courtesy Shimizu Corporation

A rapidly growing form of space tourism on Earth — zero gravity aircraft flights. Flying multiple parabolas during flight, each parabola produces 30 seconds of 0-g. Space Adventures has hosted these flights in Russia. The Zero Gravity Company will offer these types of flights in the United States in 2004.

Photo and graphic courtesy Space Adventures

This is the 2003 advertising brochure for Space Adventures. The partially completed International Space Station is seen through the window of a fictional space tourist vehicle.

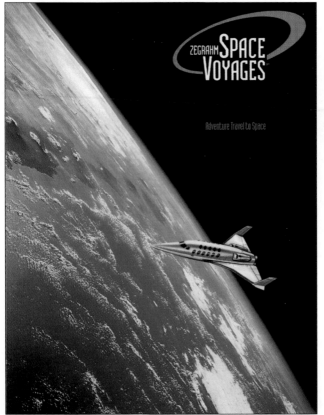

This is the cover for the Zegrahm Space Voyages brochure published in the late 1990s. Zegrahm is one of the world's premiere adventure vacation and expedition cruise operators. Their space group was acquired by Space Adventures.

These two marketing brochure covers, produced by credible companies in the adventure and expedition vacation business, made a real contribution to have the concept of space tourism taken seriously.

ASTA
AGENCY MANAGEMENT
VOLUME 69 NUMBER 1   JANUARY 2000

The Dawn Of Space Tourism

Galactic Voyages Are Closer Than You Think

Battling Burnout

Spring Skiing In Utah

Calculating Business Worth

The American Society of Travel Agents (ASTA) published the feature article titled "The Dawn of Space Tourism — Galactic Voyages Are Closer Than You Think," by Judy Jacobs in their January 2000 issue. This is *the* breakthrough article in a prestigious travel magazine that first introduced the concept of space tourism to the travel industry. The article was well researched and included interviews with most of the "Usual Suspects" (founders of the space tourism movement and industry).

Photo courtesy ASTA

The X Prize has challenged engineers and entrepreneurs from around the world to win their $10 million prize to place tourists into space. More than 25 teams are working toward the goal. This 1998 cover of the National Space Society's magazine *Ad Astra* ("To the Stars") shows many of the X Prize competitor vehicles at a rally.

Artwork by Pat Rawlings
Courtesy the X Prize Foundation
"X Prize" and "The New Race to Space"
are trademarks of the X Prize Foundation

The inaugural poster for the International Space University, 1987. ISU was founded by Dr. Peter Diamandis, Todd Hawley, and Bob Richards. The inaugural summer session was hosted at the Massachusetts Institute of Technology. The facade on the space station depicts the main entrance to MIT.

Artwork by Pat Rawlings
Courtesy Dr. Peter H. Diamandis

up to 60 miles an hour — the fastest way to cross at the time. Carrying up to 50 passengers and 40 crew, they were the high-tech queens of the sky, providing a unique and unforgettable travel experience.

The zeppelins featured private cabins and fine dining in beautiful lounges, with music and fantastic views. This form of travel ended with the start of World War II, and with the loss of the *Hindenburg* on May 6, 1937, at Lakehurst, New Jersey. Of the 97 passengers and crew on board, 62 survived. By war's end, aviation technology had advanced so fast and so far that large planes like the B-29 had captured the public's imagination. Speed was in, and the slow but luxurious airships became a thing of the past. After World War II, the ocean cruise lines experienced a similar fate, though not as devastating as that of the airships.

Today, the return of the luxury airship, or skyliner, is upon us, with several companies designing and building rigid airships for both industrial and passenger use. They have exciting plans for huge "sky cruise ships" filled with nonflammable helium rather that explosive hydrogen.

The United States military is also revisiting this form of air travel as part of a new strategic defense plan. Airships can stay on station for several weeks at a time, monitoring larger sea and land areas. The new military investment of capital and talent will surely spinoff into the commercial sector, stimulating the reemergence of commercial airships.

I see the return of the huge skyliners as not only a model for orbital tourism business development and sales, but also as a bridge between the air and orbit. In the same way sea tourism was a bridge to sky tourism, sky tourism is a bridge to space tourism.

A growing number of companies are working to create a new sky tourism industry, much as we work to create the space tourism industry. The Space Tourism Society is highly supportive of their efforts and is working to create a clear and productive interface between these complementary pioneering efforts.

The Hollywood Market

Hollywood has had a long love affair with space-themed radio, motion pictures, and television shows for almost a century. Before NASA existed and the space race began, the entertainment industry was capitalizing on and stimulating interest in space exploration and travel. A healthy symbiotic relationship exists between the real and entertainment space industries. Many reporters in both electronic and print media, both men and women, have told me that one of their favorite categories of stories to cover is space. Producers and studios know space can sell well, and seek to tap into the public's desire to experience something new.

I am one of the few people who function in both the entertainment industry and the real space development industry. I know the unique language and culture of both. Space tourism provides a powerful and profitable merger of the two.

Space people like entertainment people and their ability to reach a mass audience and to influence politicians who control the money for NASA. Entertainment people like space people because of their ability to experience truly heroic adventures and also because they are respected in society.

Many of the most powerful players in the entertainment business are supporters of space development and are slowly becoming aware of the potential of space tourism. As the power brokers in Hollywood supported the environmental movement a generation ago, I am confident many will support the space tourism movement.

A perfect example of this crossover occurred at the 1994 Academy Awards ceremony. Director Steven Spielberg (*E.T.: The Extraterrestrial*, *Jurassic Park*, and *Saving Private Ryan*) is a proponent of space exploration and a contributor to nonprofit space groups. At this ceremony, Spielberg's close friend, George Lucas, was to receive the lifetime achievement award. When the reserved Lucas was called out on stage, he was met by a standing ovation. While the creator of *Star Wars* was distracted, a huge screen dropped behind him. Spielberg asked him, and the millions watching, to turn and focus on the screen. It came to life with a live video image from the Space Shuttle's mid-deck with three astronauts floating in zero gravity. Not only that, an Oscar statue slowly spun in 0-g in front of them! The audience broke out into another round of huge applause when the shuttle commander presented the lifetime award to the normally reserved Lucas, who was brushing tears from his eyes.

The theme was repeated when the 2001 Academy Awards show began onboard the International Space Station, 250 miles above Earth. This time the astronauts opened the show and introduced its host Steve Martin, who kicked off the show's *2001: A Space Odyssey* theme.

Director and writer James Cameron (*Titanic*, *Aliens*, *The Abyss*, and the *Terminator* series) began working in 1999 on his vision of a realistic first-humans-to-Mars, made-for-television miniseries. He has received extensive assistance from NASA and the aerospace industry. My friend Buzz Aldrin and I have discussed the story, technology, vehicle, and habitat designs with him. Jim is also working with NASA to secure a trip into space, where he wants to be the first private citizen to do a spacewalk. He would also do some filming in space and use some of the footage for his Mars project.

Two-time Academy Award-winning actor and director Tom Hanks loves space. He has seen the movie *2001: A Space Odyssey* more than 30 times and, as noted above, starred in and produced major space projects. He is also highly supportive of the real NASA program and has stated publicly he wants to go. The 1995 movie, *Apollo 13*, starring Hanks and directed by Ron Howard, was the second-highest-grossing movie of that summer with more than $300 million at the box office. Again, NASA assisted in the production and allowed the crew to build movie sets inside the special astronaut training aircraft. They used flights in this aircraft to create zero gravity situations, where they filmed the actor/astronauts actually floating inside the sets just like the real astronauts.

In 1997, two competing movies about the Earth versus killer comets, *Deep Impact* and *Armageddon*, were both blockbusters, with half a billion dollars at the box office. NASA and the astronauts were portrayed in a particularly favorable light in *Deep Impact*.

In 1998 Tom Hanks produced the 12-part television miniseries *From the Earth to the Moon*. The comprehensive and popular miniseries won several prestigious awards. Hanks received extensive support from NASA, retired astronauts, and *Apollo* scientists and mission managers. Several well-known actors and directors were quoted in the entertainment trades as saying it was "an honor" to participate.

In 2000, Disney and Warner Brothers released two competing big budget and star-filled movies about the first manned missions to Mars — *Mission to Mars* and *Red Planet*.

The movie *Space Cowboys*, directed by Clint Eastwood and starring himself and Tommy Lee Jones, was also released in 2000. It was a surprise hit, playing in theaters for months. Several of my associates and I took Dennis Tito to see *Space Cowboys*. We then went to our favorite English pub for drinks and to talk about the movie, space development, and his plans to become the first space tourist. It was a wonderful evening and within six months, Dennis did become a real space cowboy.

In April of 2000, Tom Hanks was awarded the highest civilian award from NASA for his public support of the space program. The award was given to him at the 30th anniversary banquet for the *Apollo 13* crew held at the Museum of Flying in Santa Monica, California. A very grateful and emotional Hanks accepted the award, wearing a bushy beard and long hair he had grown for filming the movie *Castaway*.

The longest-running television series in history is *Star Trek* — 38 years old in 2004. According to the Paramount Studios Marketing Department, the five television series, 10 movies, and merchandise have grossed more than $2 billion. In the mid-1990s, 53 percent of Americans considered themselves *Trek* fans. *Star Trek* had a recognition factor of 90 percent among all Americans.

I knew the late Gene Roddenberry, creator of *Star Trek*. He often talked about his love and support for the real space program. That love carried over to his fans, who orchestrated a massive letter-writing campaign to President Ford to name the first Space Shuttle *Enterprise*. They succeeded in creating a beautiful crossover between space entertainment and the real space program.

Gene passed away on October 24, 1991. Because of his strong support for NASA and the positive influence of his *Star Trek* concept, his ashes were quietly flown in space aboard the Space Shuttle *Columbia* on STS-52 in October 1992.

The fifth *Star Trek* television series is called *Enterprise*. It premiered in September of 2001, with a unique opening title sequence showing the history of exploration, including images of sailing ships, flights of the Wright brothers, *Apollo* Moon images, and the assembly sequence of the International Space Station. This sequence represents the first time that the "imaginary" *Star Trek* universe and real space development met on screen.

I am 100 percent confident that in the not-too-distant future, a real spaceship will bear the proud name of *Enterprise*.

The Media Market

Mass media is our best vehicle for promoting the ideas and attractions of the space tourism industry, at least through the first decade of this new century. Only through heightened media exposure can we build a public awareness that will help boost market demand for space tourism.

We lucked out when former NASA Administrator Dan Goldin tried to block Dennis Tito's flight, causing a huge controversy. The media loves a conflict — and this was as huge as outer space! The space tourism movement and Dennis received far more media attention than if NASA had just kept quiet. Through dozens of primetime news stories and hundreds of international newspaper and magazine articles, the terms "space tourism" and "space tourist" finally entered the mainstream cultural consciousness. As has been said many times there's no such thing as bad publicity. Thanks, Dan!

I remember back in the early 1980s just how hard it was to receive any media attention, let alone serious attention, about space tourism. I remember how excited my space tourism colleagues and I would get when an article appeared. I have a collection of almost every domestic article written on space tourism through 2001. After that time, it has fortunately became too hard to keep track of all the articles and news stories.

Having media on our side, drawn by the inherent dangers and beauties of space and the real-life stories of private pioneers, will be a crucial element in reaching target markets.

Conclusion

In this chapter I addressed the question, "Is there a market for space tourism?" There are markets for both simulated space tourism, then for real suborbital and orbital space tourism when the regulatory and technological pieces fall into place.

The continued success of complementary industries, such as adventure travel, show that the mind-set is in place. This is also true due to the continued interest in entertainment products — movies, books, games, and television.

While the technology that enables suborbital access for greater numbers of private space tourists is not yet in place, we can continue to create a vision of the space tourism experience so compelling to men and women that the markets will be ready to buy when we are ready.

If we do our job well over the next few decades, then Earth orbit will become crowded with thousands of private citizens having extraordinary and life-changing experiences off world.

CHAPTER 3
THE CROWDED SKY

"We dream of orbital traffic jams."
John Spencer, June 1989

Introduction

Proponents of a space renaissance and the space tourism industry intend to crowd low Earth orbit (LEO) with a diversity of privately financed and operated off-world ventures. Managing this multilayered growth calls for a road map. I call this slowly evolving map the "Space Tourism Industry Master Development Plan."

This Master Development Plan includes significant expansion of our vision of space tourism experiences, of Earth-based infrastructure including public relations efforts; streamlining regulatory requirements; establishing space academies; building the immersive space simulation industry; modernizing existing launch complexes and developing new ones; and, of course, creating the next generation of private enterprise reusable launch vehicles and aerospaceplanes.

Any real breakthrough that dramatically lowers the cost of accessing LEO, and that increases safety by orders of magnitude, could significantly increase the number of ships and facilities off world. One critical component of the Master Development Plan that helps provide structural guidance and clarity to such a "traffic jam" is what I call an "Orbital Zoning Plan" (OZP). This chapter explores a number of elements within the OZP.

- **Places**: Every map has its own language — special terms that mark locations, provide direction, define boundaries and spaces. As the space renaissance matures, some of our early space age perceptions will need to change, as will the meanings of some of our familiar space vocabulary. For example, the term "spaceport" is used today to identify a hub of space launch activity or a proposed development here on Earth. However, what will we call ports when we begin to assemble and operate them in orbit? Reclassifying Earth-based places of space-related activity to be "Earthports" will allow us to transfer the name "spaceports" to ports actually in space. (Someday, there will be "lunarports" in orbit around the Moon. Places of lunar-related activity on the surface of the Moon will most likely be called "Moon bases.")

- **Spacecraft**: As any industry matures, specialization becomes a necessity. We will see the logical evolution of two different classes of spacecraft. The first class will be "Orbital Access Vehicles" (OAVs), which will leave Earthports and fly into LEO. These highly reusable vehicles will dock with "Orbital Transfer Hubs" (OTHs) in LEO, unload their passengers and cargo, pick up returning passengers and cargo, and immediately return to Earth. OAVs are in the transportation business using the airline business model.

The second class will be "spaceships" that are assembled and operated off world only. Many of these ships will be destinations in themselves, offering hospitality in the true cruise line business model.

- **Peacekeepers**: LEO will someday be filled with so much activity that a management and control system similar to the U.S. Coast Guard will be necessary. It could be called the "Space Guard Service" and will perform the same duties as its Earthbound cousin. Space Guard Orbital Stations will provide orbital traffic control and their fast space cutters will perform rescues, provide emergency medical services, perform ship and facility inspections, and be responsible for overall orbital security and law enforcement.

The Crowded Sky

Earth Infrastructure:
- Earthports
  - On land
  - At sea
  - In the sky
- Media
- Simulation Centers
- Entertainment/theme parks/museums
- Private corporations
- Space advocacy groups
- Factories
- Universities
- Space academies
- Government agencies
- The United Nations

Orbital Access Vehicles (OAVs):
- NASA shuttles and vehicles
- Foreign government vehicles
- Private rockets
- Aerospaceplanes
- Rail launch systems
- Laser launch systems

Orbital Service Infrastructure:
- Orbital Transfer Hubs (OTHs)
- Orbital Taxies (OTs)
- Assembly and storage areas
- Sports areas
- Receiving areas (rail launch)
- Communication and military satellites

Space Ships:
- Government spaceships
- Orbital Transfer Ships (OTSs)
- Private yachts
- Private cruise ships
- Private lunar cruise ships
- Private racing yachts

Orbital Facilities:
- Media centers
- Industrial facilities
- Private yacht clubs
- Resorts/casinos
- Sports facilities
- Spaceports
- Administrative and health centers

Scientific and Research Facilities:
- Science/observatory facilities
- Contamination facilities

Lunar:
- Orbital ports
- Lunar bases
- Lunar resort/spa/touring facilities
- Lunar sports facilities
- Science/industrial facilities

Military Facilities:
- Military bases
- Space Guard Service facilities
- Space Guard ships
- Lunar facilities

The Orbital Zoning Plan (OZP)

Imagine in the future dozens of countries, a hundred private companies, and hundreds of orbital access vehicles (OAVs), spaceships, orbital ports, assembly and manufacturing facilities, sports, and other activities all orbiting Earth at the same time, at up to 17,500 miles per hour, depending upon their orbital distance. What system can be put in place to organize these entities to operate in a safe, efficient, and fair system?

We know from studying the evolution of the shipping and aviation industries that they endured a long, hard process of developing their sea lanes and air lanes, traffic control systems, and the planning and zoning of their seaports and airports. Today they follow strict zoning and operating rules supported by international laws and enforced by government agencies. There are well-established international rules for flying with clear safe distances required between aircraft, and strict zoning codes that never allow a passenger liner to dock next to oil tankers.

The job of the OZP is to provide the same type of guidance. We will need to design and implement an OZP for off-world operations. Evolving such a plan now and over the next 20 to 30 years with international participation and endorsement, will facilitate the space renaissance and prepare the space tourism industry to handle stunning growth around the year 2030.

The main strategy is to set three imaginary spheres — or zones — encompassing the entire Earth. The first zone, called OZ-2, begins at 200 miles away from sea level. The second zone, OZ-4, is 400 miles out. The third zone, OZ-6, is 600 miles out. Each zone allows certain major functions:

- OZ-2: Orbit entry, ship assembly, reentry.
- OZ-4: Commercial, leisure, sports, educational, and residential activities.
- OZ-6: Outward lunar development and exploration.

Inside OZ-2 will orbit several Orbital Transfer Hubs (OTHs) where OAVs dock, unload their passengers and cargo, take on passengers and cargo and immediately return to Earth. OAVs never travel farther away from Earth than OZ-2. Here, space passports, security checks, and contamination issues are addressed, and everyone and everything is organized at the OTHs. Most of the crew and staff training in a 0-g environment will occur here. Orbital Taxis (OTs) travel within zones and Orbital Transfer Ships (OTSs) travel between zones.

OZ-2 also includes areas in which most of the orbital facilities and spaceships are assembled and tested. Once commissioned, these structures are then transferred into the outer zones. Orbital farms, collection areas for mass driver-launched raw materials, manufacturing, energy production facilities, Earth science, media stations, military observation, space academies, and other industrial and administrative facilities take advantage of the relatively close location to Earth of OZ-2.

OZ-4 will be the most densely populated zone. The majority of the space tourism, sports and entertainment facilities, museums, universities, cruise ships, yacht clubs and yachts, and orbital ports will be based here. This zone's lesser degree of orbital drag and decay require less reboosting (thus saving fuel and money), and offer better views of Earth, making it well suited to host these activities. This is also where government centers and hospitals will be deployed. The Space Guard Service will have its off-world headquarters and main station here.

The outer zone, OZ-6, will be the departure zone for vehicles and facilities going to the Moon and eventually to Mars and beyond. Cycling ships would traverse through this zone, dropping off their modules, and matching up with and docking with outgoing modules. Scientific and assembly stations, as well as racing vehicles, will operate in this zone.

All three zones will have well-defined orbital lanes. This allows for tighter "Orbital Traffic Control" (OTC), modeled after sea lanes. If a ship is in distress, other ships will be closer to render assistance. All zones will have Space Guard Service stations and be patrolled by fast Space Guard Cutters. There will be medical facilities in each zone, with the main orbital hospital in OZ-4.

The OZP must be designed to have limitless growth. A fourth or fifth zone could be added over the next 500 years. Sub-zones could be established to host specific purposes, along with facilities and activities we cannot yet imagine. A Polar Zone could be added to host the satellites in a polar orbit (north to south). And then there would be a zone to host craft in geosynchronous orbit at 22,000 miles above the equator. In this unique orbit, an object such as a communications satellite remains over a specific spot on Earth, orbiting at the same speed that the Earth rotates, every 24 hours. This is sometimes called a "Clarke" orbit after the famous author and scientist Arthur C. Clarke, who first explained the mechanics of these specialized orbits in 1945.

The OZP, classifications and ship naming, and the recommendation that we must form a Space Guard Service are early attempts to organize off-world activities with the long-term future in mind. Such organization will be required by the insurance and financial communities whose direct participation is critical to our success. The OZP will provide

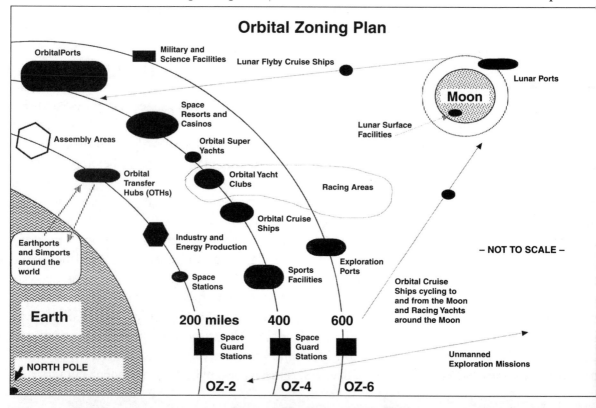

attorneys and diplomats with a framework from which to establish the required international treaties and rules for large-scale, off-world operations.

Earth Infrastructure

The rockets, space stations, space shuttles, and robotic probes you read about and see on television are just the tip of what it takes to implement the Master Development Plan. Most of the space infrastructure iceberg is hidden beneath the surface of the water.

We start with people. There are hundreds of thousands of people throughout America and in other countries at universities, research laboratories, NASA centers, aerospace companies, testing facilities, manufacturing facilities, chemical and fuel plants, launch complexes, ships at sea, tracking stations, weather services, environmental groups, media outlets, insurance companies, federal and state agencies, the financial and insurance communities, transportation companies, security firms, Congress, and inspectors inspecting inspectors who are involved daily in making the space system work.

This vast and expensive human infrastructure is one of the reasons so few countries can afford the money and talent to mount their own space programs. It also takes many decades to create such an infrastructure and experience base. This fact gives America a huge advantage in leading the world into the space tourism era. While the Russians are currently the only country offering access to space for private citizens, the United States will catch up and surpass them, just as we did during the Space Age. The big difference in the Orbital Age will be our willingness — and need — to team with the Russians, eventually bringing our space infrastructure up to speed in parallel.

This complicated system becomes even more complex and dynamic by adding to the mix the travel industry, the private enterprise-financing community, sponsorship, marketing, advertising, entertainment, real space tourists, and other industries.

The most important part of our success off world remains the professionals and staff who go. People will make or break our space tourism industry — both the simulated one on Earth and the one that expands into orbit. Their recruitment, training, and ongoing training throughout their careers are essential.

NASA and the Russian space programs do an excellent job training astronauts, cosmonauts, and ground operational personnel. There are now hundreds of space-qualified pilots, engineers, scientists, doctors, and technical experts with spaceflight experience. Most of them would love to experience space again. Many are now in the aerospace industry, universities, the military, and private enterprise. While some still do not like space tourism, they know it is coming and that it could be their ticket to ride again.

The International Space University (ISU) (*www.isunet.edu*) is one of the space community's success stories. Founded by students in the mid-1980s at MIT, it has become an international program attracting graduate students and world-class experts in many areas of space, science, and business for summer sessions held around the world. Its main campus is located in Strasbourg, France, in a new $30 million facility that opened in 2003. ISU has graduated hundreds of students with the high majority pursuing careers in space development. ISU is the closest thing to a real space academy we currently have.

There will be a need for several space academies around the world to provide Earth-based training. Military academies like West Point, the U.S. Naval Academy, and the U.S. Air Force Academy are good models to study. The Air Force Academy is already teaching a growing number of classes in space-themed subjects. They see space as their new arena of operation and responsibility.

NASA has funded some studies of the maritime academies in the United States with the goal of ascertaining which water-oriented training methods could be translated to space training in the future. Valuable information has already been derived from these studies.

Even the popular Space Camp facilities provide a valuable service by introducing young people to space, many of whom will choose careers in space because of their camp experiences. There are adult Space Camp programs that attract a wide variety of participants, from military officers to grandmothers. There are new advanced versions of "Space Adventure Camps" in development. These facilities, and the space tourism and adventure immersive simulation industry, will introduce large numbers of people to the space experience and we hope to inspire them to seek careers in space development.

Every person and every thing that goes into space starts on Earth. Building our Earth-based infrastructure to research, train, design, finance, launch, and market space is essential to our success off world.

Earthports

All real off-world experiences will begin at an Earthport. Some international airports will eventually extend their services to include access to and from LEO. These kinds of existing facilities will cater to aerospaceplanes but not rockets. Aerospaceplanes take off like conventional airliners, cruise to a high altitude, then accelerate into LEO using advanced engines. They will not pose the same safety or noise dangers of current expendable rockets.

Some existing launch complexes, such as the Kennedy Space Center in Florida, will be expanded to accommodate space passenger services. New launch facilities will include large floating platforms such as the successful Sea Launch system, which was developed and operated by Boeing and an international consortium including Russia, who supplies the rockets.

Long Beach, California, is the home port for Sea Launch, which consists of a mobile ocean oil rig named *Odyssey*, from which Russian *Zenit* rockets are launched, and the command ship, *Sea Launch Commander*, that escorts the platform to the South Pacific to launch directly from the equator. In 1999 they launched the first Russian-built rockets with communication satellites as payloads. This exciting launch approach demonstrates a new and open attitude by a major American aerospace corporation toward innovation and international team-building. It also proves that rockets can be launched from the ocean, which will greatly expand the potential for new Earthports and access to orbit. This is a significant advancement for the space tourism industry because one of the major challenges we face is the regulatory issues for launching space vehicles. If they are launched from the middle of the ocean, these issues are greatly reduced.

Several countries are encouraging other governments and private enterprise to build or expand launch complexes in their countries to take advantage of launching rockets from equatorial locations. Launching from the equator offers some significant advantages because the vehicle picks up the 1,000 miles per hour speed of the rotating Earth. This location also saves on orbital maneuvering fuels which can extend the life of a satellite or vehicle.

Half a dozen states in the United States have formed Spaceport Authorities, with the specific goal of attracting the federal government and private launch companies to their states. They hope billions of federal, state, and private dollars will flow to their state to build and operate the next generation of Earthports. Such facilities will also attract a wide variety of complementary facilities such as hotels, convention centers, retail outlets, light industry, and other kinds of support facilities — all of which provide jobs and a larger tax base for the host locations.

One must not underestimate the challenges faced by those who want to operate OAVs from the continental United States. For good reason there are many regulatory and safety issues to be considered to protect citizens and property that lie down-range of any rocket. This is one of the reasons most launch complexes are located next to the ocean. If a booster fails, it falls into the water. We were very lucky no one on the ground was killed when the Space Shuttle *Columbia* wreckage fell over a large part of Texas and nearby states in 2003. This is why the decision was made to locate all future landings of the Space Shuttle at Edwards Air Force Base in the Southern California desert.

Another challenge is that large objects such as rockets can only be moved by ocean or river barge. Being able to store rocket fuels far away from civilians is another issue. Some of the fuels used for rockets are highly toxic. Another issue is noise from rocket liftoffs, which is literally deafening. Millions of a gallons of water are used to soften the sound waves of a liftoff and to cool the launch pad. The sound suppression water can then mix with the rocket fuel exhaust to become toxic, which can harm people, animals, fish, and birds, so it must be filtered. These and other factors have great influence on the location and planning of new launch complexes.

Achieving Orbit

"Once you are in orbit you are halfway to anywhere in the Solar System."
G. Harry Stine, 1996

This quote by one of the true visionaries of private space enterprise reminds us that, once we have achieved orbit, most of the hard work is already done. That is true. Gravity is a necessary but extraordinarily expensive force from which to escape. We live on a planet with a deep gravity well — and we must fight to climb out of that well to achieve orbit.

Today there are men and women pioneering new technologies and methods to achieve orbit. Their main goals are to dramatically reduce the huge cost of escaping gravity, while increasing, by orders of magnitude, the reliability and safety of going there and coming back. There are many books available already that focus on the subject of lowering the cost to enter space, so I will touch only on the main points in this section.

Anyone who tells you it is easy to achieve orbit and that we have all the technology needed to do it now, either does not know all the issues involved or is trying to sell you stock in a private rocket company.

The United States is already far past the time we need a safer and far more economically efficient means of achieving Earth orbit. Today, there are two ways this could happen. The old way is to rely on the federal government. The new way is to rely on private enterprise. In truth, both approaches are paralleling each other and both could learn from and assist each other. There is also a third option, which is to work with international agencies and governments. The Russians have a highly successful launch system for both manned and unmanned vehicles, and other countries also have pioneers working hard on new systems.

NASA is currently studying ways to retire the current Space Shuttle fleet beginning in 2010, and to replace the shuttle with a new Crew Exploration Vehicle as part of Project Constellation. This is the name of the project to return us to the Moon and eventually land humans on Mars. One approach being considered by NASA is to develop a vehicle that carries crew, while leaving cargo and supplies to different vehicles.

The history of NASA and the major aerospace companies working to develop new ways to achieve orbit more economically and safely dates back to the National Aerospaceplane program in the 1980s, and then in 1996, when NASA awarded a $1 billion contract to the Lockheed Martin Skunk Works. Their task was to design a new, fully reusable, single-stage-to-orbit (SSTO) space launch system. The experimental program was called the X-33. If it had been successful it would have matured into the *VentureStar* program, which would have sought several billions of dollars in private and Wall Street funding to build reusable vehicles to replace the Space Shuttle fleet. NASA canceled the program in March 2001. Another program, the Orbital Space Plane, was cancelled in 2004 due to congressional scrutiny of cost and memories of past program investments with little to show.

Some aerospace companies or national space programs in other countries, including Russia, China, France, Japan, England, and India, are actively developing rocket systems and aerospaceplanes. Most are decades away from flying any type of tourist. China successfully flew its first space traveler in late 2003, onboard their own rocket with a manned spacecraft derived from the Russian *Soyuz* design. Their space agency has made public statements that China intends to land Chinese astronauts on the Moon before the year 2010. This has come as a surprise to many and a shock to the Japanese space program, provoking some of their senior officials to seek a manned Japanese space program.

The U.S. Air Force is another player in space access and development, viewing it as their new area of operation and responsibility. Some of their black (classified) research and development programs are quite interesting. Other branches of the U.S. military are also involved, mostly from an intelligence-gathering, communications, and weapons guidance perspective. But they also see a future where their personnel, bases, and ships will be operated off world.

And there are the more than two dozen private companies seeking to win the Ansari X Prize, as mentioned earlier in this book. Their specific mandate is to reach LEO in a cost-efficient manner. Their ideas and attempts, successful or not, inform the entire industry.

There is actually far more going on in the quest to achieve orbit of which the media and the public are not aware. The more real a venture is, meaning it has funding, the more under-the-radar it remains for fear of letting the competition know what they are doing or of getting premature press attention.

The aerospace industry conducts professional lobbying efforts in support of larger budgets and fewer regulations. Some private space groups and foundations also lobby for the White House and Congress to spend more money on space science and to give greater support and freedom to private enterprise in orbit. The leading citizen group is ProSpace: The Citizen's Space Lobby. ProSpace conducts an exciting program every spring called "March Storm," in which they coordinate 75 to 100 private citizens from all over the country in Washington, D.C. They meet one-on-one with congressional representatives, senators, and their staff to discuss the issues and concerns surrounding private enterprise in orbit. This has been a very successful effort in which you could participate.

There is a surprising amount of creative thought put into inventing new ways for more economical and safer access to orbit. Below are some of the most interesting approaches professionals are working on as of 2004.

Aerospaceplane: Looking much like a jetliner, the aerospaceplane would take off like an airliner from a large airport, then climb to 50,000 feet. There, a series of secondary high altitude engines would push the plane past 100,000 feet. Then, small rockets would come online to push the vehicle into LEO. The vehicle would return to Earth much like the Space Shuttle does — a horizontal landing — but it would reuse its conventional jet engines and fly to the airport. The U.S. government invested time and money in this approach during the mid- to late-1980s in a program called the National Aerospaceplane (NASP). The program was canceled in the early 1990s, but private enterprise has continued working on this important concept, winning it more attention and research funding during the past few years.

Electromagnetic Launch: Developed by the Space Studies Institute (SSI), an organization founded by Princeton physicist Gerard K. O'Neill, this system would utilize a one- to two-mile-long track that rises up the side of a mountain. Cargo capsules would be accelerated up the track at such a high speed they would literally be slung into LEO. They would then be captured and transported to their final destination.

The acceleration required creates "g" forces of up to 200 times normal gravity, so the system could only be used for inert materials such as water, fuels, and building materials. All of the energy required and the track itself remain on Earth and can be used thousands of times.

Light Ships: This is my favorite new concept and, if successful, would revolutionize everything! The U.S. Air Force has provided small amounts of funding to a very exciting concept called the Laser Launch System, which would use laser light to push human and cargo capsules into orbit. This concept was created, and is being pioneered by, Leik N. Myrabo, Ph.D., the CEO of Lightcraft Technologies. It has already been tested successfully on very small mockups. All the energy, lasers, and tracking equipment that do the pushing remain on Earth, where maintenance is far easier. The electricity to power the lasers comes from a standard electrical grid. The capsules would be relatively simple, with their own maneuvering systems and landing systems. Because the majority of the weight that conventional rockets must lift is the fuel and propulsion system required to do the lifting; freeing a spacecraft of that burden would be an enormous accomplishment. This is a breakthrough concept that deserves strong funding and support.

The Space Elevator: First introduced and popularized in science fiction stories, this bold concept has received a surprising amount of funding from NASA and other sources, as well as media attention. This explanation of the concept comes from its web pages at the Institute for Scientific Research: "The Space Elevator is a revolutionary way of getting from Earth into space — a ribbon with one end attached to Earth on a floating platform located in the equatorial Pacific Ocean and the other end in space beyond geosynchronous orbit. The Space Elevator will ferry satellites, spaceships, and pieces of space stations into space using electric lifts clamped to the ribbon, serving as a means for commerce, scientific advancement, and space exploration. Once relegated to the realm of science fiction, the concept is now being researched and prototyped by the Institute for Scientific Research, Inc.

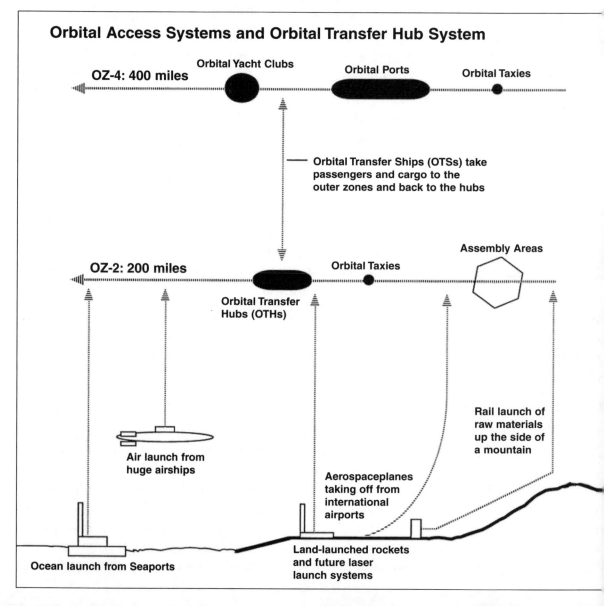

Orbital Access Systems and Orbital Transfer Hub System

OZ-4: 400 miles

Orbital Yacht Clubs

Orbital Ports

Orbital Taxies

Orbital Transfer Ships (OTSs) take passengers and cargo to the outer zones and back to the hubs

Assembly Areas

OZ-2: 200 miles

Orbital Taxies

Orbital Transfer Hubs (OTHs)

Rail launch of raw materials up the side of a mountain

Air launch from huge airships

Aerospaceplanes taking off from international airports

Ocean launch from Seaports

Land-launched rockets and future laser launch systems

SR has found that a Space Elevator capable of lifting five-ton payloads every day to all Earth orbits, the Moon, Mars, Venus, or the asteroids could be operational in 15 years."

While a fascinating concept with possible applications at the Moon, I do not feel this approach is the right one to take. I feel that, due to the single-site nature of the project, a huge multigovernment entity would need to centralize control of the project and establish restrictions that private enterprise must follow. We want this early stage of private space development to be as open and decentralized as possible — to be housed with numerous private enterprise concepts and ventures around the world. Still, the Space Elevator is a very cool concept for the future.

Huge Airship Skyports: A new generation of airships like the old *Hindenburg* are now being designed and will use nonflammable helium. A cluster of them could be used to form an Earthport, which would lift a spacecraft to 50,000 feet before release to gain speed on descent, ignite onboard engines and rockets, and blast into LEO. This approach of carrying the vehicle to high altitude prior to release was successfully used by several "X-Planes" such as the famous X-15 rocketplane that was dropped from the wing of a B-52 bomber aircraft more than 40 years ago. This approach is now being used again by a variety of new systems for launching small satellites and high-altitude experiments. The new twist is to integrate the use of an airship that can lift a much larger vehicle, then also act as a floating mission control center.

Who knows which concept will take hold first? One that both reduces the cost of access to orbit and significantly increases safety will eventually win. We must assist our allies in the orbital access field with smart and aggressive public awareness and marketing efforts. If we can establish market demand that can be taken to the bank, then they can get the funding they need to build and test their vehicles. If they do not succeed, then we will not succeed.

Vehicles and Spaceships

I forecast a clear and logical division between orbital access vehicles that take people and cargo to and from LEO, and spaceships that are assembled and operate only off world. The OAVs are in the transportation business. The spaceships are in the orbital commerce, hospitality, and exploration business.

Today, both the Space Shuttles and Russian rockets take people and cargo to the International Space Station, then return after a week or two in orbit — up to six months for the Russian *Soyuz* spacecraft. I see this system dividing into two separate but complementary industries, similar to the way in which today's airlines take most of the ocean-cruising passengers to the coast where they then board the cruise ships.

The aerospace companies are well experienced and equipped to focus on the design, testing, building, and operation of OAVs. However, a new generation of spaceship designers, assemblers, and operators must evolve who take design cues from the unique environment of space and from the ship building, yacht, and cruise line industries.

Not counting the International Space Station, *Skylab,* and *Mir*, as of 2004 there has only been one true, manned spaceship that was built and operated off world — the *Apollo* Lunar Module (LM) that successfully transported the *Apollo* astronauts down to the lunar surface, served as their base camp while they explored the Moon, and brought them back to the orbiting Command Module for the voyage home.

The LMs were brilliant successes. Designed and built by Grumman Aerospace, they operated only in the vacuum of space. They were launched into orbit inside the third stage of the mighty *Saturn V* rocket just beneath the *Apollo* Service Module. In that position, the LM was shielded from the tremendous forces of launch. It was a fragile creature that could not endure atmospheric flight or the heat of reentry, but was a marvelous vehicle, perfect for getting down to the Moon and returning astronauts to the orbiting *Apollo*.

The old pilot's saying, "if it looks good, it flies good," did not apply to the LMs. The three-story high module with its four outstretched landing legs looked like a fat spider. The first LM to fly was in fact named *Spider*. They had no smooth surfaces because they would never encounter air. Silver, black, and gold insulation foil covered most of its bizarrely-angled exterior. But, to the engineers who designed them and the astronauts whose lives depended on them, they were works of engineering and design art. For budding space architects and anyone interested in space design, you should watch the episode about the design and testing of the Lunar Module in Tom Hanks' miniseries *From the Earth to the Moon*.

Hopefully the Space Shuttle will be the only hybrid between an OAV and a spaceship that is required to operate in both the atmosphere and on orbit for up to two weeks. While the Shuttle is an incredible vehicle, it is the perfect example of a thing designed by committee trying to be everything to everyone, and not doing anything well. The vast majority of a Shuttle's life is spent on the ground enduring endless maintenance, upgrades, and testing. A single orbiter might spend only two weeks per year in orbit. The government can afford that ratio of use to downtime, but no private business could. It would be like keeping a cruise ship or airliner in port more than 95 percent of the time.

There is a real and significant difference between the design approach and operation methods for OAVs and spaceships. Quick and efficient maintenance and turnaround of cruise ships and airliners make the difference between profit and loss for the companies who own them. The same will be true for OAVs. For space stations, we have learned from real experience with the *Skylab*, *Mir*, and *Alpha* that the crew spends most of their valuable time doing repair, maintenance, supplies and equipment inventory/storage, and cleaning. Little time is able to be spent on science or just taking a break to enjoy the view. The way to reverse this is through the adoption of the OAV and spaceship systems approach, and extensive use of robotics, automation, telepresence, and other human performance enhancement technologies and methods.

Classifications and Names

We will need a logical system to classify and name the growing diversity of OAVs, spaceships, facilities, and off-world bases now in the early planning stages. By the year 2050, there could be 25 to 40 different classes of individual ships, facilities, and bases off world.

We can turn to similar situations on Earth for models and experts to help inform this work. The U.S. Navy, cruise lines, and aviation organizations can assist in this classification and naming process.

For many reasons, it is important to have a system to classify and name ships and places off world. A system is required to address insurance and finance concerns. A system is also required to assist marketing and sponsorship-acquisition efforts, as well as operations and programs to generate staff pride. We are at such an early stage of off-world industry and tourism development that the specific requests to which such a system would be applied do not yet exist. However, they are important to consider as part of the overall Master Development Plan.

While aircraft carriers and cruise ships are both large ships, they are entirely different classes of ships. One is military. One is civilian. They are financed, operated, sold, and decommissioned differently. There are dozens of different classes of modern ships. Taking their lead, I have developed classifications for spaceships and facilities in the chart in this chapter. From orbital super yachts to lunar cruise ships, each class has its own distinctive design and function.

Names are very important for the marketing, advertising, and branding industries. They are critical to creating identity and loyalty for crews of ships or staffs at places including luxury resorts. My naming scheme follows the intent of today's major cruise lines — to name its vehicles in honor and celebration of the experience. For example, the ships of Princess Cruise Lines always have the word "Princess" as part of their name — *Dawn Princess*, *Sun Princess*, and *Grand Princess*. *Princess* has come to represent a certain level of service, certain types of activities, destinations, and affluence of passengers. All navies have important traditions in the naming of their ships or shore bases. The current class of

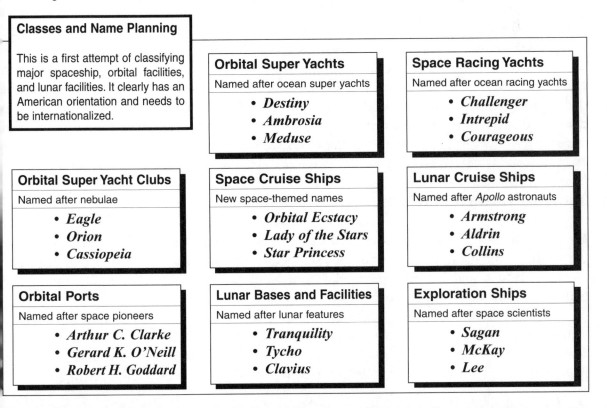

Classes and Name Planning

This is a first attempt of classifying major spaceship, orbital facilities, and lunar facilities. It clearly has an American orientation and needs to be internationalized.

Orbital Super Yachts

Named after ocean super yachts

- *Destiny*
- *Ambrosia*
- *Meduse*

Space Racing Yachts

Named after ocean racing yachts

- *Challenger*
- *Intrepid*
- *Courageous*

Orbital Super Yacht Clubs

Named after nebulae

- *Eagle*
- *Orion*
- *Cassiopeia*

Space Cruise Ships

New space-themed names

- *Orbital Ecstacy*
- *Lady of the Stars*
- *Star Princess*

Lunar Cruise Ships

Named after *Apollo* astronauts

- *Armstrong*
- *Aldrin*
- *Collins*

Orbital Ports

Named after space pioneers

- *Arthur C. Clarke*
- *Gerard K. O'Neill*
- *Robert H. Goddard*

Lunar Bases and Facilities

Named after lunar features

- *Tranquility*
- *Tycho*
- *Clavius*

Exploration Ships

Named after space scientists

- *Sagan*
- *McKay*
- *Lee*

huge nuclear-powered aircraft carriers is named for U.S. presidents including the *Lincoln* *Reagan*, and *Truman*. All Space Shuttles have names. A nationwide contest was held to name the replacement shuttle for *Challenger*. School children participated, with the winning name being *Endeavour*.

The International Space Station went through several names during its 15 years of development, beginning with *Freedom*, which became the International Space Station, and then (still unofficially) *Alpha*.

Anyone who enjoys the *Star Trek* television series, movies, and books knows one of the main characters is the starship *Enterprise*. The name *Enterprise* has a long and distinguished history beginning with a British naval ship, and two American aircraft carriers including the first nuclear-powered carrier. The first Space Shuttle built was named *Enterprise*.

The crews of each *Apollo* lunar mission were allowed to name their vehicles, which included the Command Module and the Lunar Module. They designed their own mission patch with their names and a graphic representing the main goals of the mission. The *Apollo 11* crew created a unique patch that did not have their names. Instead, it had only an eagle with an olive branch in its talons landing on the Moon. The patch represented the ethos of this mission: "We came in peace for all mankind." Individual names and individual glory were secondary to the achievement of an entire species.

The crews of Space Shuttle and International Space Station missions design their own mission patches to this day. It provides an excellent team-building exercise and is a tradition I am confident we will adopt for both the space simulation industry and the orbital space tourism industry. Attracting the interest of talented graphic and fashion designers to create them for the space tourism industry is essential. The colors of the ships, flags, logos, patches, crew uniforms, hair styles, and other identity-building symbols are important culturally, for crew and staff pride, and promotional tools.

Naming my orbital super yacht *Destiny*, which will be introduced later in this book, took a few years. The concept and design for *Destiny* started to take shape in the mid-1990s. I finally chose *Destiny* for two reasons. First, I felt I had found part of my own destiny by dedicating myself to designing and building the space yacht by the year 2030. Second, I am convinced space tourism is the key industry to opening the high frontier for humanity, and space is where we will fulfill our destiny as a species.

The Space Guard Service

By the year 2050, there could be well over a dozen countries, and hundreds of private corporations and institutes, with facilities and people in orbit and beyond. Developing quick communications and having operational control of the growing number of objects and activities in orbit is in everyone's best interest.

I have suggested the U.S. Coast Guard be the model and the initial deployers of a space-based version of themselves — a "Space Guard Service" to provide the same kind of services off world they now provide for Earth's oceans and seaports. With direct industry participation, the Space Guard should develop orbital standards for construction and operations. They would inspect and monitor spaceships and orbital facilities to ensure they abide by them. In emergencies they would provide rescue and medical services, environmental protection, and eventually law enforcement. The peaceful coexistence and mutual support of military, Space Guard, industry, science, and tourism in orbit is essential to safety, insurance interests, and investment requirements.

Having based the space tourism industry on the cruise line and super yacht models, this reinforces the rationale for using the Coast Guard as the model, the implementers, and operators of the Space Guard Service. While the smallest United States military branch, the Coast Guard (*www.USCG.mil*) has the most direct interaction with industry, science, and the general public. Their academy provides its officers with the tools to deal with a much more diverse set of circumstances and people than other military services. They are also directly involved with law enforcement and environmental protection. They have numerous shore bases, operate a fleet of ships, and control a large air force of planes and helicopters. They monitor icebergs and could develop a similar system for monitoring space debris and meteors, taking over monitoring done today by the U.S. Air Force. Their training procedures could be used to establish space academies, and their experienced personnel could become the key operators of commercial ventures once they retire from service.

I have spoken to a number of active and retired U.S. Coast Guard officers about my Space Guard Service idea. All have shown real excitement and offered to introduce me to higher officers. The U.S. Coast Guard has two of its officers in astronaut training, and they are seeking additional positions. The Space Tourism Society is actively researching and planning what would be required to actually establish the real Space Guard Service with the guidance of the U.S. Coast Guard.

Conclusion

There are more commercial, cargo, and leisure ships, and therefore civilians, on the world's oceans, than military or scientific personnel. The same is true for air travel. Clearly, someday it will be true for space commerce and tourism. New government and international agencies need to be formed to take responsibility for their safety, to establish rules and the means to control off-world activities.

Thinking ahead now, creating tools such as the Master Development Plan and the Orbital Zoning Plan, researching, planning, designing, and creating an international consensus of how to operate off world, are all essential requirements to our goal of creating a profitable space tourism industry.

PART TWO
THE SPACE EXPERIENCE

CHAPTER 4
WELCOME ABOARD DESTINY

"One expects blues and greens, whites and browns, but the pinks, purples, and yellows! I can only think about colors that one experiences seeing our planet for the first time from space."

Astronaut Byron Lichtenberg

Introduction

You are about to go off world. Through the words and thoughts of a future space tourist, you are going to get a taste for how unique and life-changing your space experience will be.

This is no small feat. Even the most articulate and insightful of today's space travelers are often frustrated in their desire to share their off-world experiences with Earth-bound audiences. Some use art and poetry; some write songs and books or create performance art shows. And some turn their tremendous talents and passions to creating space companies and societies. All of this form a passion to bring humans closer to the day when we can experience what they have already experienced so exclusively.

So here is what an off-world, week-long orbital super yacht voyage could be like around the year 2033. I am confident the experience will be even grander, more exciting, and more fantastic then described here. Welcome Aboard!

MY DATE WITH *DESTINY*

22:00, Sunday, 3.6.2033
Am here.

Still getting used to the fact I won the lottery.

Because, after all, I never win anything.

So far the Space Renaissance Corporation, owners and operators of world's first orbital super yacht, *Destiny*, are treating us first-rate. There's a lot riding on the success of this first voyage for "regular people" under their new *Orbital Ecstasy* space cruise line.

Am the only person so far who won with a ticket from that space pirate orbital cruise simulation last year. Very lucky that trip has turned out to be. Without it there'd have been no level three Simnaut status and, without that, no real chance in the lottery. Two-plus hours of 0-g experience from the aircraft flights helped in qualifying.

Am also among the first average citizens on board the world-famous

Destiny. During the past two years this queen of Earth orbit has catered to only executives, political leaders, royals, vid stars, sports celebrities, reporters, retired presidents, and their families and staff. From what I've heard, their first encounters with 0-g left some of the rich and famous stranded in the middle of volumes or bouncing off the walls like kids.

Tomorrow morning it's off to Earthport for the first leg of my week off world. There'll be a short trip up to LEO and then to the *Eagle* yacht club, where the first orbital super yacht, *Destiny*, is berthed. Once aboard there'll be the official welcome ceremony, then dinner. At 17:00 *Destiny* undocks and I'll start five full days of glorious yachting, then a day to return and reorient.

> *Note to self: Remember you're an ambassador for John Q. Public. Be on best behavior for the Space Renaissance Corporation publicity machine vid conferences on Days 4 and 5. The least you can do.*

Have met two of the other 19 guest passengers so far. Martin Smith, the retired scientist/poet and now the oldest person to venture into orbit, turns 106 during our trip. He was born in 1927 — the same year Charles Lindbergh made the first solo flight across the Atlantic Ocean. Synchronicity, huh? Also on board, the youngest passenger, Jem Klassi. Little Jamie is only seven years old. J. Klassi, her father, is *Destiny's* captain during our voyage.

Off to sleep.

07:00, Monday, 3.7.2033 (Day 1)
Am now at the Catalina Island floating Earthport, just twenty miles off the coast of Los Angeles. Shortly we launch to an orbital transfer hub aboard the Space Clipper *Conrad*, which is making its 300th flight.

> *Factoid Insert: The* Conrad *is named after the* Apollo *and* Skylab *astronaut Pete Conrad, a pioneer in the 1990s of just this kind of vertical take-off-and-landing, single-stage-to-orbit rocket vehicle. Today, the term orbital access vehicle is used for all types of vehicles that take people and cargo to and from Earth orbit. The 14th anniversary of humankind's return to the Moon will also be celebrated in a few months. The* Gaubatz, *sister ship to the* Conrad, *landed on the Moon on July 20, 2019, to mark the 50th year since* Apollo 11 *first touched down on the Moon. (The* Gaubatz *is named after visionary space enterprise pioneer and first chairman of the Space Tourism Society, Dr. Bill Gaubatz.) Her crew immediately began assembling the* Apollo *Moon Base. Weekly flights to the Moon have grown the base into the first off-world village, with more than 50 residents.*

Ran into Martin, Captain Klassi, and Jamie in the VIP lounge a few minutes ago. We were there for our first publicity session. Number one question still? "How do

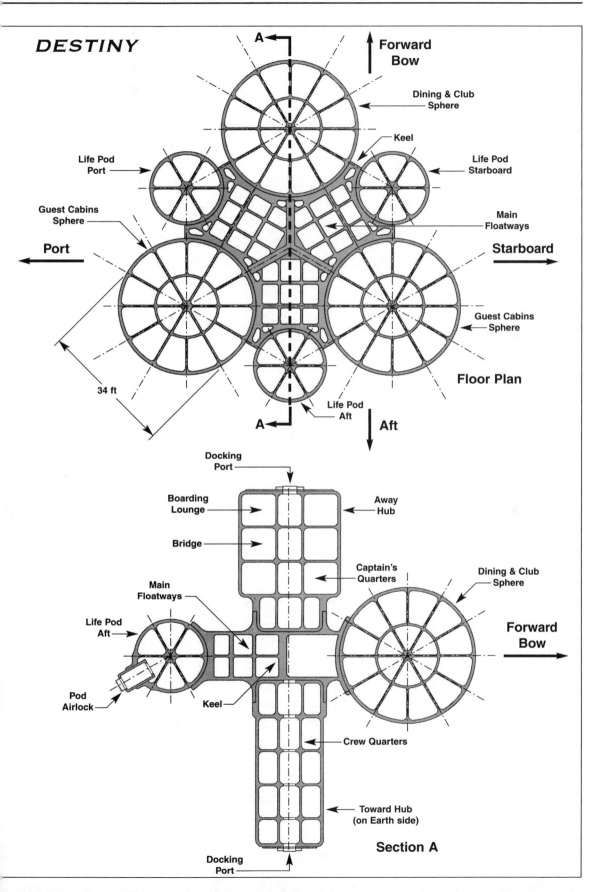

DESTINY

A

Forward
Bow

Dining & Club
Sphere

Keel

Life Pod
Port

Life Pod
Starboard

Guest Cabins
Sphere

Main
Floatways

Port

Starboard

Guest Cabins
Sphere

34 ft

Floor Plan

A

Life Pod
Aft

Aft

Docking
Port

Boarding
Lounge

Away
Hub

Bridge

Captain's
Quarters

Dining & Club
Sphere

Main
Floatways

Life Pod
Aft

Forward
Bow

Pod
Airlock

Keel

Crew Quarters

Toward Hub
(on Earth side)

Section A

Docking
Port

you go to the bathroom in zero gravity?" Geesh!

Am again very glad I did those sims over the past couple of years. Already knew a lot of the off-world language and procedures covered in the final "space-wise" briefing and emergency spacesuit demo. Some of the other passengers were clueless. Curious, but clueless. Like knowing that the word "toward" refers to the side of an orbital facility or spaceship facing the Earth, and that the word "away" refers to the side facing away from Earth. No "up" or "down." Those terms haven't been used in decades.

Did learn something new. The crew going to *Destiny* with us will serve aboard for one week, then rotate back to Earth. The Space Renaissance Corporation does this so that their specially trained and highly valuable crews can successfully deal with the health issues of prolonged 0-g and radiation exposure, and keep their families together. The happier your crew, the happier your passengers!

07:45
Just had my final medical exam. Presented my space passport to security and am cleared. I'm "good to go!"

07:50
Am aboard a hovercraft now to take us from the VIP terminal lounge out to the *Conrad*. I can see her. Wow. She's about 90 feet tall, conical shaped. Gleaming. A really pretty member of the orbital access vehicle family. She'll only take 10 minutes to get to LEO, about the same as the old Space Shuttle.

The space sims were great, but this is finally the real thing. Don't care what they say; no sim gives you this feeling. This floating, disconnected feeling you get when you are about to do something completely different. When you're standing on the edge. Lighter than air. Heart above your brain. Soul on its tip-toes. The old adage about butterflies in your stomach just doesn't do this feeling justice.

08:00
Am in the passenger module now. The *Conrad* has three, with an additional cargo level in the aft section and a cockpit for the crew in the fore section. Each passenger level has 10 comfortable flight couches, each with its own ComCenter.

I'm in the lower level and am just settling into my couch. Hold on a second — have another safety briefing. Back in a sec.

The copilot just finished the briefing and the pilot is greeting all of us right now. In a moment, the countdown. There's actually no technical need for a countdown anymore, but it's a tradition kept for the passengers' sake. A bit of nostalgic fun.

I'll be able to watch the liftoff from my couch monitor or I can watch action inside the ship via my Hyper Virtual Reality system.

And we're on our way. I chose to watch the Earth recede.

Just 30 minutes after liftoff from California, the copilot says we're passing over England. Wow. The *Conrad* continues to maneuver to a point where she can natch orbits with Orbital Transfer Hub-3, just 200 miles away from Earth.

Factoid Insert: Hub-3 is one of several LEO way stations. Vehicles like the Conrad *dock at a Hub, transfer crew, passengers, and cargo, then pick up people and cargo to backhaul their way to Earth. Each Hub is 300 feet long and 56 feet in diameter; its hexagonal frame structure hosts dozens of external hard docking ports, hard cargo modules , inflatable crew quarters, command and control, and passenger transfer and processing spheres.*

Personally, with its energy production and heat radiation panels extending toward Earth, I think the Hub looks like a Christmas tree lying on its side.

8:45

About 15 minutes away from Hub-3 now.

The copilot is telling us there's a tutorial playing for those passengers interested in how Earth orbit was organized into three imaginary spheres called "Orbital Zones" (OZs) that encircle the planet. Signing off for a sec.

Am back. Learned:

- The zones were modeled after city zoning plans that separate areas such as housing from industrial areas. There is also a system of space lanes modeled after sea lanes and air lanes. With 10 nations, 15 mega-corporations, and a growing number of private firms and even individuals having facilities and ships in Earth orbit, a plan was needed for traffic control and to guide logical expansion. Almost 100 ships and facilities are now operating in orbit and beyond. Each OZ is 200 miles farther away from Earth than the last and each specializes in different orbital activities.

- Each OZ is numbered according to its distance from Earth. The first OZ is at 200 miles, so it is OZ-2. This zone hosts most of the traffic from Earth to orbit. The hubs, orbital facilities, spaceship assembly and maintenance areas, industrial facilities, and scientific stations are also based in this zone, taking advantage of being close to Earth. A fleet of "orbital taxies" operate within zones, and the larger orbital transfer ships travel between zones.

- The second OZ, OZ-4, is the most populated, with at least 1,000 people there at all times. It holds the yachting community, residential communities, recreation areas, sports, race courses, entertainment, hospitals, media stations, administrative activities, and the soon to be opened United Nations Station.

- The off-world headquarters and research station of the Space Guard Service are located in OZ-4. Religious groups are competing to have the first church or temple off world here. Some day there will be a series of large space resorts, and the first orbital towns and ports will grow around them. The Space Renaissance Corporation will operate its fleet of orbital cruise ships in OZ-4.

- OZ-6 is just now developing and will host cycling lunar cruise ships, scientific stations, exploration, and colonization activities.

- All OZs have military facilities, spaceships, and satellites. The good news is that almost all of the early space junk from the old space age was captured as part of establishing the Orbital Zone program, or has already fallen back to Earth. So, the concerns of collisions in orbit have been significantly reduced with the Space Guard providing total traffic control and monitoring any incoming meteors, just like the Coast Guard monitors icebergs at sea.

Back on external viewing now. Can also see assembly areas for several more *Destiny*-class orbital super yachts. Passed the first Space Guard station, and the coorbiting assembly and staging area for the fleet of six Mars colony ships. They will soon be transferred to OZ-6 for more testing and training, then off to Mars. If all goes well, the first humans will step foot on the Red Planet within a year. Humanity will be a species with footholds on two planets and one moon, and on its way to becoming a true Solar System Species.

Okay, Hub-3 has just been sighted visually. We'll dock next. Then I unharness and float free for longer than I ever could before. Can't float far; still restricted to the tight volume I'm in now, but I'll be floatin' free!

08:53
Am with Martin now in the upper cabin on our way to the passenger lounge and our transfer to the orbital transfer ship.

Martin doesn't say much. At all.

I take the plunge and ask what's on his mind.

"I've never been so proud of humankind." That's what he says. He says we're an amazing race. And we've only just begun.

Met someone new. *Destiny*'s steward. Name's Tsuyoshi. He came into our passenger lounge sphere to announce our ride had docked, saying "please follow me," in a distinctly and "I thought they didn't talk like that anymore" British accent.

08:55
Very short trip into the orbital transfer ship. Am harnessed in again. Hope my luggage follows!

Now it's the second leg of the journey — our trip to OZ-4 and the *Eagle* Orbital Yacht Club where *Destiny* is docked.

This trip'll take a little longer; about three hours. I'll have more time to myself to learn about *Destiny*. I guess that'd be my only complaint so far. Not enough time between winning the lottery and whipping into space to learn all I'd like to.

10:30

Halfway to OZ-4. Just finished a tutorial on *Destiny*. Merlin was my guide. Merlin's the Artificial Intelligence Sentinel for *Destiny*. He was the first Sentinel to be granted full human rights as a sentient AI. Slightly intimidating meeting him, but I adjusted quickly.

Merlin loves his ship. I get the entire history of *Destiny's* development, assembly, and transfer into OZ-4, plus the details of her first maiden orbit a few years ago. *Destiny's* a third generation orbital super yacht. She has the classification of "super" yacht because of her size, number of passengers and crew, and luxury accommodations and services. She was assembled by linking a series of inflatable structures together, then utilized the vacuum of space to pull out the inflated structure to hold them in their predesigned shapes, creating "equilibrium structures." The internal structure of the spheres is similar to a three-dimensional spiderweb radiating out from its center. The pull of vacuum on all surfaces equalizes the stresses.

Two of the spheres contain the passenger cabins, which very comfortably accommodate 10 to 12 passengers each. The third sphere is the social sphere with all the dining, main bar, lounge/club, entertainment, and observation areas. That sphere is called the "bow" of the yacht, with the other spheres "port" and "starboard." The hub at the center of the three spheres, projecting toward and away from the keel, contains most of the operational areas, ship's bridge, gymnasiums, medical area, and crew quarters. The keel area connecting the spheres and hub contains the robotics ops areas, plant and fish farms, storage, recycling, energy production, maintenance areas, and three life pods, descendants of cruise ship life boats, docked to the keel between the spheres.

At least one skiff accompanies *Destiny* at all times, docked at either end of the hub. This mini-spaceship comfortably holds 10 people for site-seeing trips to other orbital facilities and ships.

Destiny is engine-less. She uses entirely self-contained, removable maneuvering modules, attached to the keel on both the toward and away sides. When a module's fuel is close to exhaustion, it is refueled.

Right now, I'm seeing what looks like a star, but Merlin tells me I'm actually seeing *Destiny* from her toward side. At this approach, I can see her three pearl-white inflated spheres connected to each other in a triangular form by a royal blue keel. It's the starlight bouncing off the gold trim accents and light from inside the spheres that had me thinking she was a star.

And now I see about a dozen red lights circling the yacht, which Merlin explains are protective satellites for shielding her from any meteor or space debris. Each "prosat" has a high-quality camera, and some have telescopes that can be remotely controlled by crew and passengers to observe *Destiny* from any angle, and to observe other orbital facilities and ships.

12:00

All this learning has made me hungry. (So far so good with 0-g!) I grab a lunch pack from Tsuyoshi as he floats by. My first meal off world! Nothing fancy, but tastes good and is certainly "less filling." I let a little fluid out of my drink container and play with the globules. I'm not the only one!

12:30
We'll actually be docking with the *Eagle* Orbital Yacht Club first before boarding *Destiny*.

I think twice, but agree to have Merlin provide information on the yacht club. I said I wanted to learn! He's just kicked up a real-time image of *Eagle* onto my screen. Here's an audio *<snip>* to save time:

"One of the largest facilities/ships in orbit, Eagle *is named after the Eagle Nebula, which is 7,000 light years from Earth and nicknamed the birthplace of the stars. The* Eagle *Orbital Yacht Club is the first privately financed and operated orbital port. Assembled in OZ-2 and then transferred out to OZ-4,* Eagle *was assembled by linking three, 80-foot diameter inflatable spheres together, connected by three keel sections. She has berths for six orbital yachts and provides all operational, maintenance, passenger, and crew support services.* Eagle *also hosts some of the most famous and outrageous parties and celebrations off world. Last year's Playboy Halloween Party and 0-g fashion/body-paint show are still headlining in the media and tabloids. Later this year she will host the first major film/vid festival off world."*

13:00
Eagle's blinking her docking lights in welcome. What an incredible sight. Merlin is coordinating with *Eagle's* human captain, David Weisberg, *Eagle's* Sentinel, *Seven of Nine,* and our pilot to guide us to a docking port for hard dock. Contact and a slight vibration.

We just got the "all clear" to pop our restraints and float into the *Eagle's* security lounge for our first full view of Earth from OZ-4. Back in a sec.

I'm now in the largest floating and viewing gallery in orbit — bigger than what we'll have on *Destiny*. We've got 40 feet of open volume in which to fly. Some of the other passengers have gotten themselves stranded mid-volume, but not me, or Martin, or Jamie. I've crossed the space several times already, rescuing a few flyers here and there.

Now the attraction of the view has won out over the attraction of free-flight. I'm glued to the viewport. We're all in the viewing gallery now. Mixture of reactions. Awe. Tears. Silence. Silly human grins. They've just dimmed the interior lighting and some of the exterior nav lights for our first sunrise from space. The music of the spheres fills the gallery.

14:00
Some passengers have ordered drinks and snacks from small robots who float by. Martin says he heard that the *Eagle's* crew is working now to transfer with

Destiny's old crew. Control for our voyage will then be turned over to them. We should be able to board soon.

can report success in a number of critical areas:
1) first experience using a 0-g toilet;
2) first experience communicating home from orbit;
3) extended 0-g exposure;
4) moving through 0-g with ease that's surprising even me.

The Space Renaissance Corporation publicity machine just snatched me for my first off-world interview. I was prepared to make some insightful and even snarky comments, but actually found myself at a loss for words to describe my reactions so far. They weren't surprised.

Bumped into Martin on the way back to the gallery. He said again how proud he is of humanity and how privileged he feels to have lived to experience this moment. We hear Captain Klassi's voice invite all passengers to board *Destiny*. Some of the VIP passengers look as if they know where they're going, so I'm going to float along.

5:00
I'm aboard! Since the docking port was in use for supplies and equipment transfer, we had to board *Destiny* via a skiff called *Galileo*. We undocked from *Eagle* and made a quick cross to *Destiny*'s away hub end port.

Am now in the board and departure lounge. I just thought of all of the VIPs who have passed through here over the past two years. This spot in space is one of the most famous volumes off world.

Tsuyoshi is back to welcome us aboard. A vidbot floats in his wake to record our arrival and any pithy, witty, and otherwise engaging comments. Yeah, right.

Just met my "Robotic Friend" — a ball the size of a grapefruit. My guide, confidante, and man — er, bot — Friday.

Am informed the bot is called "Nip."

We make quite the entourage on the way to my cabin. Me and Nip, Martin and his robotic friend. From the lounge we float through the hub to the very center of *Destiny*'s keel, make a 90-degree jog aft through the keel structure, and then a 90-degree jog to port into the main floatway. On the way, I'm surprised by the scent of flowers. They're all over; in floatway bubbles and in containers attached to the curved sides of the floatway. *Destiny*'s fully-functional biosphere refreshes all the air, recycles most of the required water, and provides nearly all the food from the ship's gardens and fish farms. The general lighting is bright but soft, produced by a chemical process that also mimics some of the Sun's nurturing elements.

Nip just pointed out the bright red access portal that leads to one of three life pods. Just as ships at sea are required to have enough lifeboats for all passengers and crew, so must orbital facilities and spaceships. In the unlikely event that *Destiny* must be abandoned, the pods remain attached to *Destiny* until the emergency is over, or they can detach and float nearby until the Space Guard or other spaceships arrive. Pods can also be piloted to meet rescue ships, and they have enough air, water, food, and environmental services to last two to three days on their own. Pods aren't capable of reentering the Earth's atmosphere; to attempt so in a pod would be a very bad thing.

Am following Nip into the port passenger sphere. A 90-degree jog to my left and I'm in front of my cabin. My voice is the key. Once encoded, the hatch opens and I'm home.

I can't believe it. One entire portion of the wall is clear. I immediately float over and start star-gazing again. The Moon is so close, I feel I can reach out and touch it. No portholes on these spheres. We can make sphere surfaces either partially or completely clear with just a voice command.

I'm realizing that, even 30 years into the Orbital Age, fewer than 50,000 people have been this far away from Earth. I'm going to take my spokesperson responsibility more seriously. More people must see what I'm seeing.

Factoid Insert: Destiny *cabins provide six different volumes. The living volume is the largest — nine feet long from the entry to the exterior hull, about 19 feet wide and high on the hull side, including a structural division in the center. Volumes get smaller toward the center of the sphere. There are five cabins like mine in each sphere. Each cabin radiates from the toward side to the away side, with half the volumes offering Earth views and half space views.*

I begin to explore. From the living volume I turn toward Earth and enter the bed and closet volume. And, yes, my luggage is already here, attached to the wall. The next volume includes the shower and makeup station and last, the lav. OK, back through the living volume and into the away side. There's a den/office volume and finally a storage volume.

Wait. I just heard a chime. What's that mean? Doorbell? Com call? Yes, that. "Hello," and the link opens. It's Pat Douglas, the *Destiny's* second-in-command, chief engineer, and safety officer, requesting that all passengers follow their bots to the life pods for a hands-on orientation. This is required before *Destiny* can undock from *Eagle* and begin its voyage.

I tuck behind Nip in the main floatway and meet up with the other passengers from my sphere, and a crew member. My pod is Number Two.

Orientation's over. Those things never take long. The pod's an inflatable structure like everything else around here. It's got an airlock and a lav, plus medical and survival equipment. There are 10 stations for passengers and crew, with capacity to 15.

16:45

Am unpacking now. Won't take long; only brought two pieces of 0-g luggage. Dinner and the official "welcome aboard" ceremony in the forward dining club sphere in about 15 minutes.

Off with the day-old flight coverall. Dress for *Destiny* is formal for some events, but for vid interviews, I can dress casual. I've considered the trend toward wearing only body paint, with a utility belt or arm bands for accessories. But still not sure how I feel about the exposure. Some of the paint designs are beautiful. Animal themes are the best. Humans painted as tigers, snakes, or colorful birds with wing extensions for float flying.

Merlin's on the com now and asks how I'm finding *Destiny* so far. His voice is soft, yet somehow spirited. And ageless. I'd read Merlin became self-aware in 2025 and, after choosing a male persona, named himself Merlin after the wizard of Arthurian legend. When asked why, he replied, "because it's all so magical."

I've noticed that, even though he may be carrying on simultaneous conversations with a dozen other guests or crew members and managing more than 100 service bots, Merlin always makes me feel I'm the only person he's talking to at that moment. Pretty comforting. Wish I'd had a Merlin in my corner during one of my first sims Earth-side. Actually, Nip would have been great, too.

Martin tells me later that Merlin is also in synchronous chat with half a dozen other orbital facilities, and the same number of orbital access vehicles and orbital transfer ships. And he could be dealing with several Space Renaissance Corporation executives and marketing staff while participating in a press conference, and having a private conversation with his girlfriend Jasmine, who will soon be taking over as the AI Sentinel of the second orbital super yacht, *Ambrosia*.

Merlin is also directly involved in research and design development for all the Space Renaissance facilities on Earth, and ships and facilities off world, while developing his own original ideas for future projects. His human attorney is in the process of negotiating his next two-year service contract, and he is considering offers to write a book about being the first Sentinel. Working title? *I Am Magic*. Hollywood is also calling, but he "has no current interest in becoming an actor."

17:15

Nip says it's time to go. I'm confident I know the way and ask it to stay behind.

In the floatway, I ask Merlin where he *is* aboard *Destiny*. He seems delighted to be asked.

Merlin's "awareness" is located in a shielded sphere about two feet in diameter on the bridge. He's invited me to visit him when I tour the yacht tomorrow morning after breakfast. Then he says he will "visit me in his chariot if I'd like." I laugh. An image of an AI in a Roman fighting chariot floods my mind; even to the Roman Centurion crested helmet on Merlin's sphere. And I discover that Merlin is also incredibly forgiving. He patiently explains that the chariot is a service bot equipped with a powerful wireless interface, cameras, and four robotic arms that allow Merlin to "sit in" at card games, fix things on *Destiny* and even create 3D physical art.

17:30

I did it. Found my way to the dining and club sphere. This should be very interesting. The dining and club volume is the largest, encompassing more than half of the sphere. All 20 passengers and 10 crew members, plus Merlin's chariot, fit comfortably in their four-person pods similar to dining booths in restaurants. Waist-securing belts and adjustable foot restraint bars under the arching dining table keep people stationary. Since this sphere is the bow/front of *Destiny*, the passenger pods have the best view as she orbits. The other pod zones for the crew are in the toward and away areas of the sphere.

Small service bots begin serving drinks and snacks. Printed programs in gold and silver ink tell us the ceremony, undocking, and reception will last two hours, which is just over one independent orbit. Dinner's at 19:30 hours. A break at 21:30 allows the crew to transform the dining volume into a club. They do this by deflating the pods into storage bins. Party starts at 22:00.

17:45

A voice calls for attention. It's Captain Klassi, or actually, a life-sized, 3D projection on the perimeter hull view wall of each table. After a welcome and a question to be sure we've all met Merlin, he introduces the rest of the crew. Their images wave and say, "hello."

And I get my fifteen seconds of fame! Captain Klassi introduces me as the lottery winner and spokesperson. Then he introduces Martin as this voyage's special guest. All receive a vibrant round of applause.

The captain finishes by outlining the overall voyage plan and announces that *Destiny* will log orbit number 12,000 during this trip. After undocking from *Eagle* in about 15 minutes, she will have logged 11,940 orbits as a christened orbital super yacht. Sixty more orbits will be added by the "evening" on the fifth day of this voyage.

Now comes the traditional departure ceremony. Simple, but meaningful. The captain raises his glass globe of wine and makes a toast to all the pioneers of the space age, and now the orbital age, who dedicated their lives to the pursuit of off-world exploration and development. Globes clink. (Strange to drink fine wine through a straw!) Then, with a salute, the captain promises to guide us through a safe and unforgettable voyage. *Clink*.

18:00

The captain and several of the crew leave the dining and club sphere to supervise the separation. Merlin's floating from pod to pod to introduce himself to everyone. Exterior views of both *Destiny* and *Eagle* appear on the view walls. Some passengers choose to listen to crew communications as undocking begins. Captain Klassi announces the beginning of separation. Ten seconds later we're ree. I join the others in a chorus of "huzzah." Our voyage has officially begun.

One thousand meters out from *Eagle*, *Destiny* maneuvers 90 degrees oward Earth to position *Eagle* and the planet in a receding viewpoint.

9:45

Martin and I are relaxing and enjoying the view. Martin is now not quite so silent. I hink, in a way, space gives people permission to be different than they would be on Earth. The "space" of space is a great equalizer. Those who go have shared a special experience that those who haven't can never imagine.

And we're pleased that Merlin has decided to station himself with us for the moment. I'm amazed at how quickly I've become comfortable in his presence. I consider him a friend.

Ah, hold on. A banana arrives via air mail! Tony, another passenger I met earlier, just sailed it across half the sphere to me. What a riot. Reminds me of that famous scene from *2001: A Space Odyssey* when the ape man sails a bone through the sky and it morphs into a spaceship.

The banana tastes better.

20:00

A hu-crew (human crew member) just floated over to take our dinner order. He recommends the lobster. Merlin concurs. Even so, a holographic menu appears above the table. Martin selects a salad and protein drink. I follow Merlin's advice and order the lobster, accompanied by a rich red from Sonoma.

Little Jamie just arrived! She wants to pet Merlin's chariot and she wants to ask the voyage's oldest traveler — that would be Martin — some questions. We find that Jamie is fascinated by the exploits of early space explorers, especially the *Apollo* Moonwalkers. Martin met some of the first American astronauts, and later Chinese, Russian, and Japanese space travelers.

I cup my chin in my palms and drink it in. The sights and sounds of excited chatter between the youngest and oldest persons to travel off world, and the first sentient artificial life form. I reach for the water container hovering next to my elbow and take a big sip.

A bot server just brought dinners and drinks. Jamie hugs everyone, says her goodbyes, and flies off. Merlin says he really loves that kid and reveals he may ask his AI girlfriend, Jasmine, to marry him next year and initiate a family of their own. Another thing to think about. AI marriages and families.

A coordinator of one of my Earth-side sims warned me it takes a few meals to get used to eating in 0-g. Fortunately the chefs know how to help; almost every course has some thick sauce to prevent small pieces of food from floating off. But a small tomato from Martin's salad gets away from him and levitates toward me. I reach out and snatch it from the air, popping it into my mouth. Martin winks. Bot servers quickly capture any other errant morsel in the dining sphere to prevent accidents or damage to machinery.

We all get pretty good at sneaking utensils in under each dish's covering and soon finish our meals. Martin and Merlin excuse themselves. I'm alone for the first time since leaving my cabin.

21:45

The crew's just starting to deflate the dining pods, so I'm going to get out of the way and float with "banana man" Tony over to the bar. Tony's a successful author and vid writer, also on his first space voyage. He's researching stories about the future of humanity off world. Our conversation turns to Mars, since it's now mainstream news that the colony ships are ready for testing. The last thing I want to do is get into the "Moon or Mars" debate, but I find I can't stop talking about what I've experienced during just one day off world, and that I could see myself visiting the Moon and even emigrating to Mars someday. The image of becoming a full-time "off-worlder" is not as alien as it was just 24 hours ago.

A hu-crew bartender and an old-fashioned robot bartender keep the drinks flowing. Live music filters in from the now reconfigured dining area. Time to go back for the party and the evening's entertainment!

Wow. Without the seating pods, the dining area looks and feels much larger. Dimmed lighting makes the club feel complete. Now I've really got to concentrate on stationing myself so I can see what I want to see and not float off into the middle of the volume, get stranded, and become mortally embarrassed. Some of the other passengers have already formed small clusters of three to six people who lean into each other, thereby stabilizing their group. Smart. So far about 15 people have arrived.

The music's coming from two off-duty crew members, Merlin's chariot, and two music bots. One of the male crew members is singing — no words, just lovely sounds like whale songs. Quite appropriate since we just passed over the Pacific Ocean.

Some passengers changed after dinner into colorful flowing costumes accented with lights. They're free-float dancing. It's like watching live art slowly spinning and rotating in the air. Some of them have obviously been in orbit before from the ease they are showing in 0-g. I hope I'm that good soon.

Small bot servers keep the drinks and snacks coming. But the reports were right. I'd heard that the party on the first night of a week-long voyage is a bit tame. Passengers and even crew members have to get used to 0-g. Things loosen up later in the week.

Tony just took off to make his first vid report from orbit. Am alone again. Fine by me. Am feeling a little wallflower coming on. I spend the next few minutes just watching people dancing and talking and, ack, networking? There are some business passengers on board who don't seem to be able to leave their vid headsets back in their cabins. I don't really hear what they're saying; I asked the volume sensors to mute active sound within my personal space.

The ship's band just took a well-deserved break. Merlin announces that the evening's formal entertainment is about to begin. Room lights dim and spotlights flare on a slender woman stationed in the middle section of the sphere against the erimeter wall. She is free-floating with her legs crossed and her hands in her lap. Her hair is long and red, and forms a colorful halo behind her serene face. She wears a flowing, light gray robe with no embellishments, and appears to be in her 80s or 90s. Merlin tells us we're in for a rare treat. He introduces her as one of the world's most-loved storytellers. A guest of one of the wealthy VIP scientists on the voyage, she has graciously agreed to storytell for us tonight.

Everyone in the club floats over to cluster around her. I think I just heard her say "hello" in three languages. She says this is her second time off world. The last was about three years ago, so she has had "some time to think about the meaning of it all." Therefore, she will tell two stories tonight.

Her performance begins with a few minutes of chant, in order to "take you to a place where you listen with your hearts." The first story is a favorite of off-worlders because it's all about the smells and sounds of Earth. She tells of walking through tall forests and swimming in warm lakes. I can see recognition flooding the faces of some of the passengers as she reaches the primal connections we all carry to our birthplace.

The second story comes in a smooth flow of words. It's the first time she's told this one to an audience. It's about the mysterious wonder of space travel and human destiny in the stars. At the end, she encourages us to be storytellers and to believe we all have something to share. Then she quietly slips the hood of her robe over her head and floats out of the volume. No applause. No remark. You see, in modern storytelling, it is considered a great honor when no one claps at the end of a story. It signifies that the storyteller has disappeared into the story.

And this is one of the most unexpected, yet most gratifying, offshoots of the space tourism trade: the reemergence of simple storytelling as the most admired art off world. Amidst all the instant hyper communications and information overload, people now want to hear a good story told by a real person in the intimacy of "real space" and real time.

Well, I think that's the perfect end to the day for me. I can see some other people floating away from the club area, but I'm going back to my cabin for a good night's sleep.

P.S. Just caught myself humming some of the chanting phrases we did during the storytelling. I just realize that neither Merlin or she said her name. I guess she was just the storyteller. And that, actually, was enough.

24:00

Am back home. Have told Nip to wake me at 06:00 ship's time. My first night in orbit and I'm going to be smart and follow my space-wise sim instructor's advice, sleeping attached to the surface of the volume. No free-float for me yet.

"Bed" is a sleeping bag with fancy electronics and lots of pockets. Free-float sleeping is fine once you are comfortable with longer duration 0-g. I've learned that waking up in free fall is a bit startling because you see your arms floating out in front of you but you don't feel the strain of holding them out.

Factoid Insert: In 0-g the human body comfortably and naturally assumes the neutral body position in which the spine elongates up to three inches due to not being compressed by gravity. This causes a curvature of the posture as well as the raising of the arms.

I've been in this position most of the day, as has everyone else I've seen.

Whoa, what was that?

I just noticed a device entering my sleeping volume. And there's another! Oh, right, I know what these are. I saw them on *Eagle*, too. They're air flow compensators or "bugs" to most people. They work to keep the air supply fresh and healthful. They're basically scrubbers. In 0-g, air doesn't flow the way it does in normal gravity. It's critical to scrub the air in volumes where humans live in order to keep carbon dioxide exhalations at an acceptable level. A bug's sophisticated sensors identify a human or other air-breathing creature in a volume and move into position to push fresh air in their direction.

Bugs also provide a form of mobile environmental control by heating or cooling a volume's air on request. The number of bugs matches the amount of activity in a volume. They can even change color to match the surface to which they attach themselves. They're as invisible and quiet as technology can make them. Bugs allow *Destiny* to not require air ducts, thus greatly simplifying her design. The bugs glide over to me and reposition to keep working through the night. By listening very hard, I can barely hear them move. Otherwise my cabin is completely silent.

After zipping into my bag bed, I voice my volume to "black." The bugs remind me of something I read about *Destiny*. She's got no electrical wiring or centralized plumbing. All 400 to 500 electrical devices, ranging from hair dryers and shavers to master computer systems, are battery powered. All water/waste management systems are independently serviced by bots. Batteries today last just about forever and are the size of old coin money. The service bots just replace them when needed. The art and science of recycling water and waste is so advanced that most of the water units are rarely serviced. All drinking water is always fresh and fortified to keep us healthy. OK, enough detail.

Hanging in my sack with only me, myself, and I, I can't help thinking about certain night activities including 0-g lovemaking. There are a lot of old jokes about "rendezvous and docking." Funny thing is that all the early speculation, and even research about sex in 0-g, was just talk. People are people and will do it anywhere they are and any way they can. In 0-g people just hang on to each other or some surface and have fun. By voice command I could order my bed/bag to double in size to accommodate another person.

06:00, Tuesday, 3.8.2033 (Day 2)
Nip's beeping. I'm up!

Dreamt last night. Of floating in space, believe it or not. So waking up was really easy.

I've got about 15 minutes to exercise with one eye on the news that's playing on the view wall, and the other watching Earth.

Schedule is light today. No spokesperson responsibilities. I'm off to explore and make more friends.

12:00

Just had one of the most fascinating experiences off world — watching my laundry being done in my cabin. *Destiny* is filled with innovative technologies, but equally, if not more, significant to her success are breakthrough operational concepts. "Laundry bots" are a perfect example. Instead of a fixed laundry volume, *Destiny* offers four mobile laundries composed of three box-shaped service bots about 30-inches on a side that float from cabin to cabin doing laundry while the passenger or crew member is absent. The nickname for several bots following each other in formation is "bot train."

I'd heard from others that this relatively mundane task was something to see, so I arranged to be in my cabin when the laundry bots arrived. Entering one at a time, they moved to the sleeping volume, hard docked with each other, forming a laundry machine almost eight feet long.

Accompanying the laundry bots were two smaller, spherical service bots, each 24-inches in diameter, with four human-sized and configured arms and hands. They got the laundry from my laundry bag and loaded the center bot. Once the hatch was closed, a special cleaning gel from the left side filled the washer and the clothing was washed, then pushed into the remaining bot on the right side and dried. The bots then folded the clean, dry clothing and placed it into a fabric dresser composed of many different-sized, closable, see-through pockets. The laundry machine then disassembled and left the cabin. The dirty gel will be recycled like just almost everything else onboard and in orbit.

Factoid Insert: This innovative system of "Mobile Ship Functions" is just the tip of the iceberg for many operational strategies that shrank the required physical size of Destiny *by almost 30 percent, significantly reducing her construction and operational cost.* Destiny*'s creator and famous space architect calls these innovations "Interactive Architecture," or IA.*

Most of the volumes in *Destiny* are used for multiple functions. For example, the passenger boarding and departing lounge also functions as the co-ed gym, classroom, space float preparation, suit storage, and medical center. Basic surgery, as well as delivering babies, can be done in the medical center. However, for normal health issues there are three "Mobile Medical Centers" with six standard med bots. Each med center is typically stationed in the life pods but, like the laundry bots, will move to cabins to serve people with illnesses or minor injuries.

The ships skiffs are also used for several functions, as well as several other volumes onboard including the life pods.

Passengers and crew can order items from the mobile store, which are then immediately delivered to their location. The beauty parlor caters to women and men and is usually sent to a passenger's cabin, or to one of the other volumes to host groups.

There are few fixed storage areas on *Destiny*. Instead, dozens of mobile storage bots of different sizes and functions, some with refrigeration, are stationed in volumes, including floatways not being used by people. "Courier bots" take the storage bots to where their contents are needed. This system is also used for the fish and plant farms, as well as recycling and general maintenance services. All together there are close to 100 service bots being moved and stationed throughout *Destiny* at all times. Trash is also managed with this system.

Passengers and crew hardly ever see this continuous dance of bots. Merlin and the crew always know where everyone and everything is inside and outside *Destiny*. All bots, crew members, and passengers have transmitters for location tracking. Humans can also be tracked by individual heat signatures and by health monitors in the health system network.

Another reason that IA works so well is that *Destiny* was designed with a double primary floatway circulation system. This triangular-configured system through the spheres and the keel allows most of the movement and stationing of the bots to be invisible to the passengers. The away-side floatway and the toward-side floatway parallel each other. Passengers are required to use the away-side, while crew and bots use the toward-side. 0-g is another plus in this system as bots can be stacked and stationed in a wide variety of ways, leaving clear openings and passageways for the crew and other bots.

Other operational systems, supported by Merlin and the crew, allow *Destiny* to operate in the most efficient and safest manner possible. Merlin and the crew can spend most of their time personally tending to the passengers' needs and interacting with them, making their off-world experience as enjoyable and stimulating as possible.

13:00

Out to shop and collect art.

Many early space entrepreneurs were surprised that personal experiences and artwork became the first big businesses off world. Not 0-g manufacturing, energy production, or even mining the Moon and asteroids. Those industries are only now being established and should boost the second generation of space tourism. For the past three decades, the private sector has grown and thrived on providing the space experience to humans, and on selling and buying stories, space art, poetry, music, images, vids, and fashions created off world.

The hunt for space art treasures and the promotion of the artists has become a big thrill for the rich and famous, professional collectors, galleries, and museums. In space, a growing number of technically-trained personnel, as well as administrators, chefs, and service workers, have demonstrated wonderful creative ifts. Being on orbit seems to stimulate creative juices, proving that there truly is a pace renaissance in full bloom off world. Space and design academies can take ome of the credit because of their wise decision requiring training that balances echnology, history, social studies, and the arts.

Zero-gravity and low-gravity sculpture is "in." Jewelry fashioned from lunar rystals or rocks is "out." What people want is to be able to touch, wear, or view omething created in space. That's the selling point, no matter the item.

The first space news story I viewed this morning was actually about the ew "Art World Colony" being assembled in OZ-2. It'll be operational in three nonths and will provide part-time residency to 100 artists, designers, poets, omputer scientists, and vid writers from around the world. Merlin said he'll be ecturing there on an ongoing basis.

7:00, Wednesday, 3.9.2033 (Day 3)

Can't believe it's day three already. As my log attests, the environment is really aking hold. I can't seem to be bothered to make log entries throughout the day. ust enjoying the opportunity to relax.

But today we're back on schedule a bit. And have a very exciting trip lanned.

Since I'm an *Ecstasy* lottery winner and spokesperson, I get to spend most f today away from *Destiny*, traveling about in the skiff *Galileo*, with Martin and a ew other passengers. Our mission? To return to OZ-2 and tour the assembly area vhere the world's first orbital cruise ship, *Orbital Ecstasy*, is coming to life. And do nore vid interviews!

8:00

Passing through the keel, we make a 90-degree turn away, pass the bridge and ay hello to Captain Klassi. Maria Rodriguez, *Destiny's* head chef, will be the pilot or today's adventure. Almost all crew members have at least two onboard ositions, and some are qualified and experienced in up to five. For example, Rodriguez is also second engineer on all biosphere operations, with a specialty in ish and hydroponic farming.

Our flight plan's been confirmed with Space Guard.

We rave about last night's dinner while settling in. Can't hurt to butter up the pilot! We're shown the location of emergency spacesuits in a storage area etween the airlock and small lav. "In the unlikely event of loss of atmosphere, don he suit immediately. Each suit contains a ten-hour supply of oxygen." The mplication is that the problem is either fixed or we're rescued before ten hours are up!

Galileo accommodates 10 people, including the pilot, and can be operated ntirely by remote control from *Destiny* or by the Space Guard. No copilot necessary. Like *Destiny*, *Galileo* is an inflated sphere. However, she is elongated o maximize viewing surfaces since her main purpose is sightseeing. Her entire hull is transparent from the inside, but opaque and white from the outside. She has active Sun glare filters, which also assist in keeping the inside temperature quite comfortable. Typically she utilizes four maneuvering modules attached to her away and toward sides, but since we're going to and back from another orbital zone, two more modules were added during the night. Each module is preprogrammed with the flight plan.

There is no fixed control station or "bridge" on board, which took a little getting used to. The pilot actually just wears a belt with a localized guidance computer and power supply; he or she uses a VR helmet to review information and issue commands

We're strapped in. Systems are "go." Separation has begun.

"Three — two — one." A slight vibration and *Galileo* is now drifting away, a free-floating spaceship.

Thrusters kick in. The trip is uneventful, but filled with gorgeous views of dark space away from Earth, and our home planet on the toward side.

11:00

We're at OZ-2! The *Ecstasy* assembly area is in sight. (Actually there is no fixed assembly dock like the dry docks used to build ships on Earth; it's just a point in orbit.)

We watch as several specialized assembly and material ships crews work under the direction of an assembly office ship. *Ecstasy* will be so large that a new orbital transfer hub was assembled first to house all of the assembly crews. She'll accommodate 200 passengers and crew members for week-long orbital cruises. Within five years, some of her sister ships will provide lunar cruises.

As Rodriguez maneuvers us around the assembly area, Merlin provides a running commentary. It's clear these two have done this before!

Here's what I learn:

- *Orbital Ecstasy* is modeled after the *Eagle* super yacht club with its inflated spheres 80-feet in diameter. However, *Ecstasy* has 10 spheres, while *Eagle* has only three. *Ecstasy*'s spheres have a slight pinkish tone. Her life pods are stationed between the spheres with her dozens of maneuvering modules and four skiffs partly recessed on each side to keep her profile elegantly smooth.

- *Ecstasy* assembly is on schedule. She begins trial orbits in one month. Her maiden orbit is already being promoted as the most spectacular, star-studded event ever off world. A "Who's Who" of high society will be aboard, making history and headlines. I'll be there, spending my second week in orbit. I understand now why they are cranking me through the publicity machine during my time on *Destiny*. I'll be a space veteran and much better spokesperson by the time I get to *Ecstasy*.

- *Ecstasy* will cruise 24/7. She will never dock with a port. She is both a cruise ship and a port unto herself, with constant service from a fleet of support ships and orbital transfer ships that move passengers and supplies.

Rodriguez just stationed *Galileo* directly above *Ecstasy* for a bird's-eye view of her beautiful and logical design. Her forward section has three spheres in a triangular formation. The most forward sphere is for parties and for Earth and space viewing; next inward on the port side is the dining sphere; the one on the starboard side is the theater, entertainment, and sports sphere. The next three spheres are aligned in a row with the center one actually being two spheres — the first on the away side for luxury cabins, the second on the toward side for the ship's bridge, operations, security center, telepresence center, and main docking ports.

The final three aft spheres are in a triangular formation and contain all the remaining functions including the gym and spa, ship stores, game and sim center for the kids (of all ages), the space walk ports, health center, as well as crew and entertainment staff quarters. A double ring of protection satellites will escort *Ecstasy* at all times.

Her cruise schedule is based on a one-week main cruise for the majority of the passengers. The *Ecstasy* will run 112 orbits with 16 orbits a day at 90 minutes in orbit. However she offers several cruise schedule options. Some passengers stay only for three days, or 48 orbits. Others for five days, or 80 orbits. Some people can stay for a month. The more orbits, the greater your bragging rights.

When a seven-day main cruise ends, *Ecstasy* will rendezvous with a cluster of special orbital transfer ships who have passengers for the next cruise. They will then take the current passengers and some crew members back to OZ-2 and the orbital transfer hubs. This process is designed to minimize the time required to complete and start cruises, maximizing revenue-producing time on orbit.

Merlin just noted that *Ecstasy* will be the first off-world ship to have two Sentinels directing ship's operations. The two AIs consider themselves brothers, and are completing their training at the Space Academy located on Christmas Island in the South Pacific, Space Renaissance's main Earthport. Next week they transfer to *Ecstasy* to begin final training under Merlin's guidance.

Well, we're just a few minutes out from docking with the assembly area headquarters ship. That's where I'll do my promotional interview. Afterward, it's back to *Destiny* and some time to myself. Yea.

08:45, Thursday, 3.10.2033 (Day 4)
Today's the day I became "aware."
Let me explain.
During the past 70-plus years of space exploration and lunar development, a naturally evolving phenomenon called the "Space Awareness Movement" (SAM), grew into the equivalent of an off-world religion. Those practicing it never call it a religion, however, because it focuses on enhancing and broadening the self and

the cosmic awareness of the individual.

Today, millions of people practice the awareness. There are hundreds of awareness centers on Earth where the Earth-bound can have high-quality, simulated space awareness experiences. The lucky who have gone off-world can refresh themselves with large beautiful views of Earth and space in awareness sessions conducted by guides who have extensive off-world experience.

One of the most popular off-world tours of the rapidly growing space tourism industry is the "Seeker Tour," designed for participants who are seeking the awareness. Attracting citizens of all ages, these small group tours include a master awareness guide and at least one week off world.

So, at Tsuyoshi's invitation, I'm at an early morning awareness session in the dining club sphere's private away lounge. Turns out he is older than he looks and is an awareness master, level six. Impressive. There are nine levels of awareness. A guide must have been off world three times to advance to the next level.

The only other person here is Sally Root, the security director for one of the VIPs aboard *Destiny*. This is her fourth trip off world. Apparently, Tsuyoshi feels that both Sally and I appear stressed. A typical 90-minute, one-orbit session usually does wonders.

I've always been curious about the awareness movement, but haven't gone much further than that. I thought the only way to truly understand it would be to go off world. So now I have no excuse.

All awareness sessions begin with the Sun setting. This provides a calmness and soft lighting, allowing the participants to "disappear" into their own thoughts. Tsuyoshi begins with light drum beats. He's singing a song now about the wonders of the universe, the beauty of Earth, and about Earth/Gaia sending her children outward, furthering her sphere of life and awareness. The rhythmic beating of the drum and his humming is very primal.

Then there is 30 minutes of silent floating and, if one desires, looking at space or Earth or just floating with closed eyes singing to oneself.

Tsuyoshi's new song becomes one of humans working together in an inspired quest for peaceful exploration and knowledge, about dreams of becoming a Solar System Species and then, someday, a species of deep space.

Suddenly, he stops. We drink in the view in silence, aware of connections, vast distances, no distances, commonalities, and differences. Unity.

The Sun rises. He has told the Sun screens to allow in more light, so the volume is incredibly bright. The birthing of a higher awareness.

Wow I really feel relaxed and inspired. Sally is actually bouncing around like a kid. Tsuyoshi is stretching in ways I did not know humans could. Just wonderful.

Later in the dining sphere over tea, Tsuyoshi explains that, while we consider ourselves modern humans, we retain the genetic imprint of cave dwellers and are far more primal than most of us will admit. The awareness encourages us to embrace that primal core and merge it with a modern awareness of the universe, ourselves, and advancing technologies.

Time to process and absorb the most intensely internal day I've experienced so far. The drum beats echo through my mind as I float back to my cabin. I will check out the awareness movement when I get back on Earth.

3:00

A few words on "zooming" and "pinging."

In the old days, astronauts were required to exercise several hours every day to prevent muscle and bone atrophy in 0-g. The longer the mission, the more important was the daily exercise.

Today, however, off-world exercise is done for fun and for personal fitness. The "bone-breaking" effects of 0-g were mitigated long ago.

Destiny offers many options for play, inside as well as outside the ship. For example the floatway inside the toward hub is five-feet in diameter and 25-feet long — the perfect place to bounce off the walls, stressing legs and arms, and sharpening reaction time during space ball games. Racing around the triangular main floatway between the three spheres definitely works up a sweat. Everyone uses the toward floatway as an obstacle course by dodging the never-ending stream of service bots.

"Zooming" and "pinging" has become the first recognized off-world sport (besides yacht racing) with its own zoom teams and zoom stars, major sponsorships, and arenas called pits. "Pits" is a hold-over term from when the sport was created a decade ago by bored assembly workers who, when management wasn't looking, would zoom from one assembly strut or surface to another using only muscle power and 0-g skills. Spacesuits back then were not only indestructible, but offered a wide range of arm, leg, and torso movement. Whenever someone missed their mark or was "bounced" by a fellow zoomer, he or she would have to use jet packs to maneuver away from a fiery reentry to the pit, accompanied by much hooting and laughing. Small and large amounts of personal credits changed hands based on who had the longest zooms and most pings, from surface to surface.

Today, three teams of five players play in well-lit pits with visitor stations all around. Like a game of 3D basketball, the goal is to put a grapefruit-sized glowing ball into a bin.

It's like the old Rollerball game, but without the blood! Speed, accuracy, and the ability to work as a team, are what make winners. Combined male and female teams are the best, with males launching the females who then throw the ball. Most of the game stars are women.

10:00, Friday, 3.11.2033 (Day 5)

I'm on my way to my very first space float.

Actually, nobody "walks" in space. One walks on the Moon. Or bounces. But one floats in space.

Merlin will be my coach. I'll be floating solo. I'm looking forward to the quiet time with nothing but my own breathing to keep me company.

Destination? The away end of *Destiny's* hub. I'll be going "out" through *Galileo's* airlock.

Spacesuits today actually remind me of the old wet suits that divers used. The helmet, while elegant in design, still contains all of the mechanical systems for regulating air flow, temperature, and communications, as well as two cameras. Part of the helmet extends halfway down my back; it contains the rebreather processing unit. I've got 24 hours of air.

A maneuvering belt with mini-thrusters provides total mobility. I can initiate maneuvers by voice command, manually, or by selecting a preset course and just enjoying the ride. If anything goes wrong, Merlin or a human safety officer will send a *Destiny* crew member out to reel me back in.

Merlin keeps the final-check chit-chat at a minimum.

"Have a good float!"

I'm in the airlock. The outer hatch is opening. I launch out into the void.

I'm my own little spaceship.

But I can only go as far as the "fence" created by the orbits of the protective satellites. This is for my own protection. I ask the suit to stop about 200 meters away from *Destiny*. Rotating 90 degrees, I'm now facing deep space. No Moon or Earth in view, just me and the universe.

This is the experience I've dreamed about my entire life. I'm living it right now.

How can I share this?

Time for some snaps. Deep space.

A two-second jet burst and I can get some great shots of *Destiny* and Earth. We are over the Pacific Ocean again. That background almost makes *Destiny* look like a ship at sea.

After about 60 minutes, I realize it's time to go back in. Plus I have to pee. Merlin confirms he grabbed a few snaps of me zooming in my spacesuit. No one can tell it's me, but that really doesn't matter.

Feet first and legs straight, I slide right into *Galileo's* airlock, and the hatch closes behind me.

And Martin's here, waiting to help me out of my suit and to ask lots of questions! He's going out again after lunch.

19:00, Saturday, 3.12.2033 (Day 6)

All good things come to an end. As do all fantastic, awe-inspiring, life-changing, and incredibly unbelievably good things.

After five extraordinary days of orbital yachting, it's time to head home.

It's the last dinner aboard, and as is tradition, any passenger or crew member can make a toast at the last dinner. I've actually been working on one for several days, but it kept changing with every new experience.

It's finally my turn. Facing all my yacht mates, I raise my drinking container and thank everyone for a great experience, especially my hosts, the Space

Renaissance Corporation. I recall what the first private space traveler, Dennis Tito, said more than 30 years ago when he first floated aboard the old International Space Station, "I love space." I modify it to "I love the people of space." Cheers all around, and the party goes into the wee hours of the night.

At one point, Captain Klassi floats over to tell me he enjoyed my toast and that I'm welcome aboard his ship anytime. Nice to know I now have an off-world home, too.

I know I'll never be a space explorer like those brilliant and brave men, women, and AIs who will soon be exploring and settling Mars, then the inner planets, the Asteroid Belt, and the moons of Jupiter.

But I do have the right now to call myself a space traveler.

I wonder if I'll live long enough to witness the beginnings of the Star Explorer era. With life spans approaching 150+ years, perhaps!

08:00, Sunday, 3.13.2033 (Day 7)
Last morning. Last breakfast. A bit sad.

Goodbyes all around to yacht mates and crew.

We're watching the docking between *Destiny* and *Eagle* right now.

Everything's over except reentry. Ugh.

Back to the cabin for one last check and a farewell to Nip.

I'm in the first group to board *Galileo* for the short jog to *Eagle*.

Once aboard *Eagle*, the public relations staff escorts me to the viewing gallery for my final vid interview off world. She never seemed small when I was aboard her, but *Destiny* is dwarfed by *Eagle*.

I see the next group of *Destiny* passengers tumble into the gallery. I feel like a space veteran and decide to act like one, helping a few newbies get to flatholds on the sides of the gallery. Some even recognize me from the vids I've been doing for Space Renaissance!

We've just been told an orbital transfer ship is ready to take me and most of the other passengers from my voyage to OZ-2 and Hub-6. Waiting for this stage of the journey is the hardest part. I know it's not rational to be afraid of reentry, but accidents do happen. I'll be on board the latest aerospaceplane back to Earth. It'll be just like an airliner, landing at Los Angeles International Earthport.

We undock from the Hub and begin the fall toward Earth. Despite super materials and active sound-canceling measures, a roaring sound grows as we encounter the atmosphere. As the deep pink color increases outside the viewports, all I can think of is the mythic tale of Icarus, the boy who flew too close to the Sun.

I want to look away or close my eyes, but I'm fascinated by the fiery show. Gradually, the glow and the roaring dissipates. I feel the engines spool to life. We become a graceful supersonic airliner.

And, hey, gravity's back! I feel it asserting itself once again on my spine, arms, and legs.

A smooth landing. I'm back on Terra Firma.

The copilot reminds everyone to remain seated at first, and only begin to slowly stand up when we can support ourselves while holding on to something. It will take a little time to get used to gravity again.

Before I've reached for anything, my wrist communicator starts to chirp.

Yep, I'm back on terra firma!

Conclusion

I hope you enjoyed your vicarious off-world yachting experience. E-mail me at *JSSDesign@aol.com* and tell me what you think.

I believe that something close to the orbital yachting experience scenario just described will happen. However, I believe it will be even more fantastic than I have been able to forecast.

Clearly I am mixing futurism with space development, along with my own vision and plan. It is critical to look at any long-term, 20- to 30-year development plan as a future forecast in which you are lucky to be even 25 percent accurate. But striving to articulate such an experience now serves to identify the many areas of important research and development which must be undertaken regardless of the final outcome.

For those of you interested in design, the next chapter explores the creative process, design, and technical issues behind the creation and design of *Destiny* and *Eagle*. One concept that this chapter illustrates is that people, me included, who are not aerospace engineers or scientists, have a meaningful role to play in creating an off-world future.

CHAPTER 5
DESTINY DESIGN

> "Simplest solutions and do that again and
> again until it all looks so obviously
> simple that everyone will say 'anyone
> could design that,' and they will never
> know what you went through, how much
> God went through, before developing his
> hydrogen atoms and blades of grass and
> eggs."
>
> _Buckminster Fuller, April 18, 1974_

Introduction

The quote above, from one of Buckminster Fuller's lectures, is very meaningful to me in my profession as a conceptual designer. Fuller was the design scientist who originated the significant concept of "Spaceship Earth" back in the late 1950s, and was the inventor of the geodesic dome. His quote has provided the foundation for my own design philosophy since the mid-1970s. I strongly believe that simplicity is the key to the successful design, financing, and safe operations of off-world facilities and spaceships.

The orbital super yacht _Destiny_ is a dream ship. Today she only exists in thought, drawings, and models. But I can visualize floating through her floatways as she carries me in orbit 400 miles out from the Earth. I'll admit that some nights I dream about her. She has become my companion and my challenge — a demanding mistress always calling for more attention and detail. Every year she comes into greater focus. She may be the most advanced work in future-oriented space architecture existing today.

My ambition is to complete her architectural design and operational strategy by the year 2020; then inspire a major corporation to finance her detailed engineering design, construction, and assembly; and to be onboard for her maiden orbit around the year 2030.

I created _Destiny_ in 1997 and have been researching and designing her on a part-time basis since. I expect to devote the next 30 years of my career to bringing her to reality. I also expect to have a great time doing it, meeting extraordinary people and extraordinary companies along the way. I already have. Ultimately, it is not only the technology that will bring _Destiny_ to life, but the inspiration, creativity, and dedication of many people, including possibly you.

The Design Frontier

Since the mid-1990s I have been using the phrase "exploring the design frontier" to describe my design career. This just seemed a natural statement that simplified and streamlined media interviews and presentations.

I believe designers can also be explorers, exploring the "design frontier" with the same spirit and dedication as those who explore places on Earth and beyond. Designers can make exciting design discoveries and have the pleasure of naming them.

I did not set out to explore the design frontier. I have always been a designer from early childhood, and a curious person interested in both history and the future. As a teenager, I was fascinated with the *Apollo* lunar program, while at the same time studying architecture. I was influenced by the movie *2001: A Space Odyssey*, the original *Star Trek* television series, and space- and future-themed attractions at Disneyland. I started studying the technologies for designing homes of the future, spaceships, and Moon bases. I have never stopped.

I love to sit down with a blank piece of paper and go on a "design adventure." You never know what you will come up with during a few hours of sketching and writing. And there are those few precious moments when you know deep down that you have originated a new perception, concept, or design that makes all the years of study, hard work, and sacrifice worth it. You have brought something new to the world, and you hope the world will be better for it.

I celebrated my first quarter century as a space architect in September 2003. Since 1978, while still in architecture school, I began researching, designing, experimenting, and exploring off-world architecture. Two projects for which I completed design studies have flown in space — the International Space Station, for which I received two design awards from NASA, and the Spacehab module, which flies in the cargo bay of a Space Shuttle. Fewer than half a dozen architects have been pioneering the emerging field of space architecture as long as I have. It has been a fun and exciting 25 years!

The most fascinating aspect of off-world architecture is the presence of zero- and low-gravity environments. Designing for such conditions is real pioneering work. All unique environments — under water, in the air — we struggle to design for the one thing in common — gravity. Off-world architecture changes everything. It forces you to challenge all of your normal assumptions of up and down. People can fly in zero gravity, and so can you, as a designer. This is truly thinking way outside the box *because there is no box.* Space provides a place for true three-dimensional design.

When you conceive of a design that no one you are aware of has conceived before, you are exploring the design frontier. There are no guidelines or pathways, just your own instinct, experience, and open mind to rely upon. Having solid scientific and engineering knowledge about the environment for which you are designing is essential, or all you are doing is fantasy design. The environments and sets that production designers and special effects teams develop for movies and television look good and look real but are, in the end, just fantasy. Exploring the design frontier requires actually building and operating designs in extreme environments, with peoples' lives at stake.

I am not an aerospace engineer, but I work with or review my designs with experienced aerospace engineers and astronauts. I find that not being an engineer is a great advantage when your purpose is exploring entirely new design concepts. The engineering and details can always be accomplished after the concept is established.

Many times the questions you ask are more important than their answers because they lead you in new design directions with opportunities for innovative and even breakthrough designs. Seeking the "wow" solution is a wonderful and empowering way to approach design.

For example, I was completing my architectural master thesis back in 1981 at the Southern California Institute of Architecture, while teaching a design class to help cover my tuition. The class was designing a Moon base for 100 people, with an emphasis on making the base as self-sufficient as possible. One quiet student came in one day with a sketch for the habitats using bamboo as the structural support system. Yes, bamboo. She had transferred from a botany program at a university and theorized she could bring the bamboo seeds to the Moon and grow the stocks in nitrated lunar soil. They would grow more than a foot every day and, in the low lunar gravity (0.16-g), easily support multi-story habitats. The bamboo stalks could be tied together with ropes in a way similar to what is done in Asia for scaffolding on construction projects. Bamboo trusses could also be constructed to serve as beams or other horizontal supports.

Everyone in the class, especially me, was just blown away by her approach — her "wow" solution. Only someone from outside the space field could have discovered this entirely new and innovative approach to lunar architecture. She received an "A" for the class.

Regardless of whether you are designing a Moon base, an Earth-based family home, or a Mars-themed resort and spa simulation venture, you always have the best chance for success by designing from the inside out. Start with one human being, then a couple, then a group of people. Strive to create as flexible and beautiful a stage for human affairs as possible and remember that technology changes rapidly, but basic human needs and desires are much the same today as they were thousands of years ago. They will remain much the same a hundred years from now whether you are on Earth or off world.

Destiny Design

I believe *Destiny* will be a third-generation orbital yacht and the first orbital super yacht. The first generation would include such craft as the International Space Station, which could be the first space yacht if she is purchased and converted by private enterprise or government. I believe the second generation of space yacht will be a forerunner to *Destiny* that is smaller and simpler, but still with inflatable spheres expanding from a hard shell containing all the environmental life support, energy production, and docking systems. The shell would be a module that would fit into the Space Shuttle cargo bay, with the spheres deployed from both ends after the module is placed in orbit. It could also be launched on an expendable booster. One or two rescue vehicles would be added after deployment.

Modeling *Destiny* after ocean-going super yachts made her overall design quite easy. Super yachts have guest cabins, dining and party areas, bars, a gym, a diving area, lifeboats, crew quarters, engineering areas, and a bridge. So does *Destiny*, although some of her volumes are used for multiple purposes to reduce her size.

If you excuse the expression, one cannot design an orbital super yacht in a vacuum.

I realized almost at once I was not designing just a spaceship but a comprehensive space system. Questions beyond *Destiny*'s design and operations required answers which in turn had real influence on her design. Where would she be assembled and at what altitude would she orbit? Answers to those questions led to creating the orbital zoning plan, defining the clear distinction between orbital access vehicles and spaceships, and creating the design for the orbital transfer hub system. The question of where to dock for services and exchange of crew and passengers led to the creation and design of the *Eagle* orbital yacht club and the realization of the need to establish the Space Guard Service.

Since day one of creating *Destiny*'s design, I intended to use inflatable spheres and structures as her main parts. This decision led me to the issue of utilities, which led to the breakthrough idea of projecting that long-life batteries for electrical devices and independent water and waste management stations would be feasible, and that they could be serviced by a small army of service bots. Determining who would manage the complicated scheduling and overall operations led to my research into artificial intelligence (AI) and telepresence, which led to creation of the AI sentinel and Merlin concepts. All of which led to the interactive architecture (IA) design approach and my struggle to make *Destiny* as simple a design as possible.

There are detailed design drawings and computer graphic images of *Destiny* in the color sections of this book. You might want to review them before proceeding with this chapter. By 2005, we will have a website that reports on *Destiny*'s development progress. More and more people from different professions are beginning to work on her, bringing their own talents and ideas to the design process. She is an excellent platform from which to ask questions, test ideas, and to learn. A rallying point with a name, shape, and context.

Interactive Architecture (IA)

IA is the overall name I created to describe the many design and operational strategies used to make *Destiny* as simple and flexible as possible. IA allowed the reduction in physical size of *Destiny* by almost 30 percent, which would significantly reduce her construction and operational costs.

The essence of IA is simplicity. Taking a cue from Buckminster Fuller's quote at the beginning of this chapter, I think of most of my architectural designs as beginning with an empty stage. This is similar to how Fuller thought of his geodesic dome as the ultimate in flexible buildings.

With this approach to architecture, the people come to the stage bringing their goals, furniture, equipment, communications, tools, and supplies. Then they interact. The more flexible the stage, the more options the people have to use it, and the easier it is to accommodate change in technology and attitudes.

Today's rapidly changing communications technologies are giving people the opportunity to be more connected, yet more independent. The constrictions of location and time are becoming meaningless. Cell phones, wireless laptops, the Internet, video conferences, and other technologies and attitudes make hotel rooms, yachts, airplanes, trains, cars, restaurants, and even backyards fully functional workplaces. Living rooms ecome Internet cafes, entertainment, and home shopping centers. People are working, earning, and playing 24/7 across nations and across cultures.

We have more control and more flexibility than ever before, and these trends are accelerating. People already live in interactive architectures more than they realize. Most will accept, even demand, more interactive environments in the future, making *Destiny*'s unique design and operational strategies easily acceptable.

Destiny's assembly process begins as deflated structures, brought into orbit with no equipment, but which are then deployed using the pull of vacuum, then they are connected together. Based on this approach, the following design and operational strategies were created.

What will make these design strategies work is zero gravity, which allows the total three-dimensional use of the volumes. It would allow the movement and stationing of many service bots through the volumes without conflicting with passengers and crew members.

The most important design breakthrough is to have no fixed utilities or single power distribution system. All of the hundreds of devices requiring electrical power use long-lasting batteries, which are replaced by service bots. All devices requiring water, or dealing with human and other waste, are self-contained, requiring no general plumbing and also serviced by bots.

Air flow, heating, and cooling are accomplished by small mobile units nicknamed "bugs," which position themselves around people and remain as quiet and camouflaged as possible, circulating warm or cool air.

Mobile Ship Functions (MSFs) are another key design strategy. Several traditional ship functions that used to require fixed spaces, but that are used a comparatively small amount of time, such as the laundry, medical center, ship's store, and gym have no fixed location in *Destiny*. Instead, the specialized equipment or supplies required for the function are mobile, going where and when they are needed. For example, the beauty parlor and stylist move to a passenger's cabin for individual or group sessions. The ship's doctor and mobile health center move to a cabin unless there is serious injury or illness. The ship offers a virtual store through which passengers can order items online for delivery to their cabins.

Scheduling the use of volumes, activities, travel time, stationing of bots/modules and people would be key to efficient use of the volumes, and the passengers and crew time. Deploying more than 100 service bots inside and outside the ship, dozens of telepresence operators on Earth or in orbit, and 10 human crew members 24/7 will be the main responsibility of the ships artificial intelligence sentinel (AIS).

Like a three-dimensional chess game, with moves anticipated far in advance, the AIS and crew members would strive to seamlessly and invisibly match passengers' needs and desires with the required resources, equipment, and crew expertise. A primary key to the success of this complicated real-time dance is using wireless transmitters, along with a highly sophisticated sensing system inside and outside the ship to track everyone and everything at all times.

While some of these designs and operational strategies may seem strange now, I firmly believe that in 20 to 30 years they will be commonplace, and possibly surpassed by smarter strategies and technologies that make IA even more interactive and invisible.

Zero Gravity

Destiny is a zero gravity ship — one of the first spaceships designed with 0-g as a main design driver and as a wonderful advantage.

I started designing zero gravity facilities and spaceships in the late 1970s, when space architecture and zero gravity design were new and exciting. It was very hard to break free of the bonds of Earth gravity-based thinking and design. After many years of experimentation with some interesting methods, I finally became comfortable designing for zero gravity. Now I feel restricted when designing in normal Earth gravity!

There are real challenges dealing with 0-g. Space sickness is an issue that is being addressed and should become a nonissue by the time large numbers of people are traveling off world. More serious than temporary sickness is the loss of bone strength and muscle mass due to 0-g. Without normal gravity to stress our bones and muscles, they begin to lose calcium and start to atrophy. This is not a major problem for someone spending only a few weeks or a month off world, especially with a proper exercise regimen. However, for long-duration flights, or for multiple flights, it is a real concern for the crews. There is much research going on to counteract these negative physical effects.

Some think the solution for the long-term negative effects of zero gravity is to create artificial gravity by spinning structures in space. This is a 1950s and 1960s concept created when scientists and doctors had no knowledge about the physiological effects of 0-g on humans. It is why early science fiction and science fact movies showed wheeled space stations like the one in *2001: A Space Odyssey*. They slowly rotated, producing artificial gravity through centrifugal acceleration. However, spinning a large structure is a brute force approach to solving the health issues of zero gravity.

To have artificial gravity, you first have to invent it. The costs of the structure, systems, docking stabilizers, and many other elements will make it impossible to finance and a nightmare to operate. There are also physiological issues to be considered for people living in spinning environments.

Zero gravity is free and a primary reason people want to go to space. A better approach is to learn how to live with 0-g, not do away with it. I am confident there will not be any wheeled space stations or other configurations of spinning space stations and ships.

Robots, Telepresence, and Artificial Intelligence (AI)

One of the most important capabilities we must have for a profitable off-world tourism industry is to be able to extensively use robots, telepresence, automation, and AI systems.

There has been a ridiculous debate between scientists, engineers, and others over humans versus robots for exploration and the use of space. As the talk show host Jay Leno would say, "oh, shaddup!" Robots are tools that enhance human capabilities, reduce operations costs, and allow for work in dangerous environments.

There are several industries already successfully using an increasing array of robots and remotely-operated vehicles (ROVs). Oil, mining, military, scientific, treasure-seeking and entertainment industries have been funding research, testing, and deploying of robots and ROVs for the past few decades. Their investments provide models from which we can learn technologies and experiences to apply to our off-world ventures.

Deep-sea construction work is one excellent example. In some ways deep-sea work is more difficult and dangerous than space work due to the changing water pressure and, like space, the life-threatening cold. Salt water is also more corrosive than the space environment.

The most interesting film ever shot around and inside the *HMS Titanic* was for James Cameron's blockbuster film *Titanic*. He utilized both manned deep submersible vehicles (DSVs) and unmanned ROVs. Using special effects money and new technology, they developed new mini-ROVs and the software to control them.

The use of remote-controlled military drones for surveillance, and even for attacks, is the approach for today's high-tech U.S. military. Their successful use in Middle East conflicts has demonstrated their effectiveness and provided an ability to operate in hostile territory without endangering pilots or troops.

The only off-world development and operational strategy that makes economic sense is combining humans and robots as fully integrated teams. NASA has done a good job at pioneering research into human/machine interface systems. One of the most exciting and innovative systems is called Robonaut, which is a robot in the shape of a human from the waist up with head, torso, and two arms with extremely flexible hands. Its camera eyes provide depth of field vision for the human operator controlling the robot hands. The goal is for the human operator of Robonaut to be inside the International Space Station while the robot works outside. This system is planned to deploy in several years with the goal of significantly reducing the amount of time humans must spend conducting dangerous spacewalks.

It is clear we must design space structures that utilize robots and telepresence for assembly on orbit and to maintain them over their 30- to 50-year operational life. An assembly system with a ground crew using space-based robots and a small, three- to five-person orbital crew to service and repair the robots must be used if construction costs are to be kept within a reasonable window.

Orbital facilities and spaceships will utilize extensive automation systems. This is a trend seen for several years with ocean cruise ships. These ocean ships are run by two different staffs: the command/engineering staff who run the physical ship, and the hospitality staff who cater to passengers. The trend has been to reduce the engineering staff through automation and increase the hospitality staff, therefore providing the passengers with more services and attention. Our orbital ships will follow and advance this trend. For example, out of a 10-person professional crew, three would hold command/engineering positions, with the remaining seven tasked to hospitality.

The use of mobile robots will also be part of the unique experience aboard orbital ships. Floating robots called "bots" can carry food, drink, and other items and services to passenger cabins and serve them in the public areas. They will have their own AI system onboard to be able to interface directly with passengers. The bots will have methods of recognizing the passengers, possibly through voice recognition or cards they carry that emit an identity code. Bots will know passengers by name and special preferences. They can accept tips at the end of their cruise that are shared with human crew members. All of the ships will have highly sophisticated master guest computers with which the passengers can interface to control cabin temperature, lighting, and many other comfort issues.

Passengers can form enjoyable relationships with these AI systems, making their voyages more interesting and personalized. They may even be able to continue a virtual relationship with the AI systems from Earth and look forward to visiting with them again on their next space cruise.

My research shows that by the year 2010, humans will have advanced computer science to the point that it will advance itself at an accelerating rate. Computers will be designing and programming computers. The old science fiction theme of computers becoming self-aware will no longer be fiction but fact.

I developed the term "sentinel" for the AI entity within our orbital ships because their main responsibility will always be for the safety of humans, both crew and passengers. I believe it will be a natural evolution for the AIs to choose their own personalities and names.

The AIs will not be affected by zero or low-gravity. They can be shielded from radiation, allowing them to serve for years at a time. Sensors and bots will be their eyes and ears, allowing them to monitor and maintain all systems on or off ship. The human command crew will have the ultimate authority over the AIs, within clear restrictions that could never endanger the passengers or other ships. It is likely that the human ground and space crews, and the AI crews, will develop a strong working relationship and even friendships.

The Outer Space Design Environment

In order to design facilities and ships in the space environment, you need to have a basic understanding of that environment. I love science, and have been studying space science since I was a kid. That made my transition to space architecture pretty easy.

According to internationally agreed upon scientific definitions, outer space begins 62 miles (100 kilometers) away from Earth. This altitude equals 328,100 feet, or a little more than 11 times the height of Mt. Everest. One cannot orbit the Earth at 62 miles because there is still too much atmosphere causing drag. Even 100 miles away is too low for a sustained orbit — 200 miles away is a good orbit that will require some reboosting but is accessible to the Space Shuttle and Russian rockets.

In addition to moving away from Earth, you must achieve the incredible horizontal velocity of Mach 25 (this is 25 times the speed of sound). This velocity equates to 17,500 miles an hour, and without that velocity you would quickly fall back to Earth. (The speed of sound — Mach 1 — is 761 miles an hour at sea level. However, the Mach 1 speed changes at higher altitude, based on several factors, such as air pressure and density.)

At orbital velocity you would travel from the west coast of the United States to the east coast in just 10 minutes. The fastest military jet fighters or spy planes (that I know of) can only travel short distances between three and four times the speed of sound. The fastest human aircraft was the X-15 rocketplane that flew Mach 6.7 (4,520 mph) in October 1967, and that flight lasted only about eight minutes!

The heat a space vehicle must endure on its reentry to Earth is at least 3,000 degrees.

If the vehicle's heat protection system fails the way it did on the Space Shuttle *Columbia*, the vehicle and crew will be destroyed. With our current technology, it is not at all easy to travel 100 miles away from the Earth.

Space is not the most hostile environment in which humans live. Visiting or living deep under water is more dangerous and unpredictable than space. In fact, space is a very stable environment. Other than meteors or solar flares, there are no other natural surprises in Earth orbit or near space. There are no earthquakes, floods, water waves, exterior fires, winds, tornadoes, volcanoes, hurricanes, mud slides, snow loads, animals, or bugs. It is very hard to achieve Earth orbit, so it limits who gets there, and once there, you can see anything coming at you from a long way off. One can have excellent security in Earth orbit and outer space.

Some of the unique characteristics of outer space are:

- Vacuum
- Zero gravity
- Radiation and solar flares
- Temperature differential
- Solar glare
- Meteors and orbital debris
- Orbital mechanics
- Space sickness
- Air flow and liquid movement
- Contamination

This book is not an engineering text, so I will cover these space characteristics and issues in very broad strokes just to provide you with a basic understanding of our space tourist environment.

Vacuum: More than 99 percent of the known universe is vacuum. Nothing. Nada. Zip. Vacuum wants to pull things apart — to suck the air out of a spaceship or person. An unprotected human in vacuum would lose consciousness in about 30 seconds. After that, bad things start happening such as boiling blood, exploding lungs, and freezing. Death can come in just a few minutes.

However, vacuum is not all bad. In a vacuum there is no friction such as that caused by air. You can reach incredible velocities in a vacuum. It can be used in some manufacturing processes and in inflatable structural systems with the vacuum force actually deploying the structure. Vacuum should be thought of as an important natural resource. I like to compare vacuum to trees used in building log cabins. Early settlers found the trees of North America to be an abundant natural resource for building homes. I am confident that many new ways to use the vacuum of space will be discovered and invented.

Zero gravity: (See previous section.)

Radiation and solar flares: The Earth's atmosphere shields all life against most of the harmful effects of solar and cosmic radiation. Humans evolved in this protective shield and must take that shielding with them when they leave Earth's biosphere. Radiation is highly quantifiable. Its short-term health effects on space travelers and other living systems are well understood.

Proper radiation shielding will protect a space tourist for one week to one month off world. The main health concern is for the crews who will accumulate years of space time. Their long-term exposure could put a cap on the amount of time they can spend off world without receiving harmful radiation exposure.

Future owners and operators of space facilities and ships will be forced to seek innovative solutions for crew radiation protection. They will not be able to afford the loss of crew expertise. Solutions will also be required before exploration missions to Mars and other planets in the inner solar system can ever take place.

Solar flares are a real and present danger for space tourists and crew members alike. Huge eruptions on the Sun's surface release extreme amounts of harmful radiation that reach Earth in just less than eight minutes. An astronaut doing a spacewalk might not have enough time to get inside before receiving a harmful dose of radiation. Even crews inside the Space Shuttle face possible dangers because their vehicles are not as well shielded as the Space Station.

Scientists have become far better at monitoring and even predicting the weather on the surface of the Sun, but like earthquakes and tornadoes, the science of flare prediction is still in its infancy.

Temperature differential: There is no temperature in a vacuum; you must have some sort of material to react to heat. Heat is only present if you are near a star or some other radiating source. At Earth's distance from the Sun (93 million miles), the side of a spaceship or spacesuit facing the Sun will be at +250°F, while the side facing away will be approximately –250°F. A 500-degree differential between sunlight and shade means that each side can change in seconds if an astronaut turns around. This is also true if you are on the Moon because there is no atmosphere to conduct heat. Materials and fabrics must absorb heat slowly and dissipate it the same way, or else the rapid and extreme heating and cooling will destroy the structure or spacesuit.

Heat is difficult to disperse in space because the vacuum provides no medium such as air or water through which it can travel. Our current technology requires the use of heavy and expensive heat radiating panels, which are attached to the International Space Station or the Space Shuttle. New technologies and methods are needed to deal with heat dispersion or better yet, to recycle it as part of the energy production and management system.

Solar glare: Light from the Sun is extremely brilliant in space and on the Moon because there is no atmosphere available to act as a filter. The glare can be dangerous to any unshielded human eye. Cockpit windows, viewports, and space helmet faceplates must be shielded — instruments and sensors must also be protected as a sunrise happens every 45 minutes, 16 times a day, in Earth orbit. The Sun is so bright, astronauts usually use blindfolds in order to maintain sleep through the multiple sunrises and sunsets experienced while in orbit.

Meteors and orbital debris: Meteors travel in space. Meteorites crash on Earth. Meteors are a natural hazard off world — rocks can range in size from sand and dust particles to miles across. They travel at incredible velocities from all directions. Some are light in composition like dirt, while others are dense like iron. A dense meteor the size of a green pea would easily penetrate a spacesuit, probably killing the occupant. One the size of a marble could cause real damage to the Shuttle or the Space Station. One the size of a baseball could destroy a spacecraft.

We track meteors and space junk the size of baseballs and larger, and warn the Shuttle to get out of their way. Unfortunately, there are thousands of pieces of space debris in Earth orbit. They are pieces of early rockets and satellites, lost tools and bolts, and even trash, all of which can turn deadly at orbital velocity. All debris on orbit will eventually fall to Earth, but in the meantime it poses a real threat. In the future, laws will prevent most of the space debris, but there will always be meteors.

International Space Station modules are wrapped with layered blankets like bulletproof vests to stop the marble-sized objects. In the future, super-strong materials must be invented to protect spaceships and facilities. Active defense systems will be needed to deflect or destroy dangerous objects. These could be small satellites that surround ships and are deployed to catch or deflect dangerous incoming objects.

A new business is being born from the job of cleaning up orbit. Some companies have been formed to devise ways to clean up the mess we have already made of LEO. They have some innovative and interesting ideas for space garbage collection.

Orbital mechanics: The mechanics of orbits can drive you insane because they are counter-intuitive. In orbit at 200 miles altitude, you will be traveling at 17,500 mph. However, to maintain a Clarke orbit 22,000 miles away from Earth, so that you stay over a single spot on the planet by orbiting once every 24 hours, you will be moving at only a leisurely 6,000 mph. On the other hand, if you don't want to maintain orbit, but instead want to leave the pull of Earth's gravity and escape to a place like the Moon or Mars, you must attain at least 25,000 mph. This is what the *Apollo* astronauts had to achieve in order to go to the Moon. To return to Earth, you must change your orbit so that your new orbit intersects the planet's atmosphere, where drag takes over to slow you down for a proper landing.

To rendezvous and dock with another spaceship, you must first match orbits, which means to achieve the same distance and speed away from Earth. After maneuvering closer and closer to the docking port, a craft must then be lined up perfectly so that position and angle of the vehicle is perfectly matched with the docking port along its centerline. Once all is in sync, the incoming craft slowly moves forward until the two docking connectors lock onto each other. The pilot then checks for a "hard dock" and equalizes the pressure between the two vehicles.

Mass is a big part of orbital mechanics. The most massive part of a spaceship or facility will always naturally gravitate toward (face) the center of the Earth. The aft section of the Space Shuttle with its main engines, orbital maneuvering system, tail fin, and other heavy structure and equipment, is the most massive part of the vehicle. It wants to face the Earth, causing the nose of the Shuttle to face outward. This phenomenon is known as a "gravity gradient orientation." To prevent this, the Shuttle reaction control thrusters automatically fire to keep the vehicle in the desired orientation, usually with the payload bay facing toward Earth so that science instruments can do their work and astronauts can take beautiful photographs.

Space sickness: Approximately half of all people, regardless of background or training, currently end up with some symptoms of space sickness while on orbit. Symptoms are very similar to sea sickness, although there is no direct correlation between the two afflictions. In space it is easy to get disoriented because of 0-g. Walls may become floors and a former ceiling may become a wall or floor. The inner ear takes time to get used to this new situation, and thus, nausea can result. No one can predict who will or will not get sick until it happens. Some of the best test pilots get sick in orbit while some scientists with no flight experience do not. Fortunately, almost everyone adapts to the 0-g environment within one or two days. We are learning how to reduce occurrences of space sickness by experimenting with a variety of drugs, bio feedback, and other methods.

Air flow and liquids: Zero gravity creates unique air flow and cleaning problems. Without gravity, any particles of food, skin, and dirt, as well as hair and fluids, will not fall to the floor because there is no floor. Everything eventually collects at air duct intakes and must be vacuumed away. Astronauts must be very careful not to fall asleep in an area with little air flow. Without proper air flow, the carbon dioxide from their exhaled breath can accumulate and actually smother a person. Liquids must also be carefully controlled or they can escape and form spheres that just float around. This is fun with orange juice, but not with vomit or liquids from the waste management system. Fluids and electronics do not mix well.

Contamination: In the closed life support systems of spaceships, the need to clean everywhere is critical to avoid long-term contamination from biological sources. Some of the problems the Russians had to deal with on their *Mir* Space Station were molds and fungi growing in unreachable places. They would often spend hours scrubbing surfaces, but were still unable to completely reach the affected areas. This is why the interior racks of the International Space Station were designed to pivot outward for easier accessibility. Cleaning, contamination control, air flow, and accessibility to all interior surfaces are vitally important considerations in any off-world design project.

Key Enabling Technologies

This section identifies a half dozen key technologies for which research and development programs should begin soon. I believe all these technologies are needed to enable the economic and safe development and deployment of orbital facilities and ships in the future.

You may be surprised by some of my choices, but keep in mind that my design approach is to simplify every ship system and its overall operations in order to reduce cost, enhance safety, and simplify the integration of advanced technologies.

My continuing study of major technical trends and future forecasting was critical in identifying the enabling technologies listed here. I am confident there will be new technologies coming online in 20 to 25 years that will make the assembly and operation of *Eagle* and *Destiny* easier and more economical.

The key enablers for Interactive Architecture are:

- Inflatable space structures
- Long-life batteries
- Self-contained water/waste systems
- Self-sustaining biospheres

- Air flow compensators (bugs)
- Maneuvering modules (MMs)
- Nuclear power and fusion

Inflatable space structures: Humans have always used the natural environment to build our structures. Island people build with bamboo and palms. Log cabins arose from an abundant supply of local trees. However, there is nothing similarly available in Earth orbit.

Or is there? The vacuum, energy from the Sun, temperature differentials, and zero gravity are all natural elements in the orbital environment we can learn to use to create an entirely new form of space architecture.

Nature tells us that habitable space structures should be spheres. Spheres are the perfect form for space structures because they are equal in all dimensions and are entirely three-dimensional. Because the vacuum of space pulls on an object equally on every square inch, a spherical shape will be in equilibrium if the entire surface is of the same strength, and if its interior structure works as a spider web to counteract the expanding forces. The sphere remains in equilibrium — stable with little or no structural stress — giving it a long life in the space environment. This is why I call these structures "equilibrium structures."

There are many other issues to consider, such as the stress caused by the extremely wide temperature differences between the Sun facing side and dark side of a space structure, space debris and meteorites, as well as radiation and other degenerative elements. Spheres present a large surface area that can be slowed down in LEO by impacts with atmospheric particles, causing the structures to require reboosting. However, I feel to assemble large structures in orbit by connecting a series of spheres still makes the most sense. If a sphere deflates due to an impact or other accident, passengers and crew can simply escape to another sphere. An equilibrium structure does not have to be a sphere. Any shape with the correct interior structure can be in equilibrium.

Other advantages of inflatable structures are that they can be made and tested on Earth. Deflated and packaged in small volumes, they can be brought into orbit and then inflated to create a large habitable volume. Inflatables weigh less than hard structures and the skins can be designed to self-seal if they are punctured.

A skin can be developed that has the ability to collect solar energy like solar panels, then migrate that energy to receiving/storage stations. Internal heat energy generated by people and equipment can also be recycled in the same fashion. The large surface area of a sphere facing away from the Sun can be used to radiate excess heat. Skins can also be created that are clear from the inside and opaque from the outside like one-way glass. They would provide automatic Sun glare shielding and could, on verbal command, become more opaque, like curtains, to shield a room or area.

In terms of architectural appearance, I feel that spheres are the perfect form for the zero gravity environment, and dome structures feel right for the surface of the Moon and Mars. To me, a series of spheres connected together simply looks like a space structure.

Clusters of spheres provide many architectural options and look like they belong in a zero gravity environment — just like a long ship with a pointed bow or a jetliner with wings looks and feels right in the water or air.

Inflatable structures could be manufactured to display different colors, graphic designs — even glow from the inside. The design potential with inflated space structures is limitless. Rather than "think outside the box," why not "think outside the sphere" instead?

Long-life batteries: If you were to ask for my advice on a single technology in which to invest that will have significant and positive influence on making the off-world tourism industry profitable, I would answer "batteries." Yes, batteries. Those ever more powerful, long-lasting, smaller, and lightweight hearts of laptops, cell phones, tools, medical devices, wrist watches, toys, lights, and now smart clothing. But what do batteries have to do with orbital facilities and spaceships? Possibly their very existence would make these facilities financially viable private enterprise ventures.

The distribution of electricity from a central source through buildings and ships causes major complexities and costs. Most fires in buildings are caused by electrical problems. If we could reduce or remove electrical wiring by having the hundreds of electrical devices self-powered, we would significantly reduce facility and ship mass and cost, while enhancing safety. *Destiny* is designed without wiring.

Simple concept, but I consider this approach a real design discovery. It evolved out of my self-imposed question of how to simplify spaceship design. During the past decade battery technology has made tremendous advances — striving for smaller, longer-lasting, lighter, and cheaper batteries for a growing host of devices and industries. Every one of the three to four hundred electrical devices, from hair dryers to lights, stoves to shavers, and communication devices to projection systems, will be powered by batteries. All communications, sensors, and operations will be wireless. This design approach makes connecting the spheres, hubs, and life pods to the keel far simpler.

The batteries (or whatever they may be called 20-plus years from now) will be monitored by the ship's AIS. When batteries run low, they will be replaced and taken for recharging by small, floating service bots.

I have described this approach to several engineers and astronauts who are very familiar with the Space Shuttle and International Space Station. All have been surprised by this concept and design approach at first. After reviewing the *Destiny* design, they start getting excited that this is, in fact, a breakthrough approach. It can be used in Earth-based buildings, ocean ships, and so on.

Self-contained water/waste systems: *Destiny* is also designed without plumbing. Conventional ship-wide plumbing would add a significant amount of mass, structural penetrations, complexity, and assembly time. Following the same approach as the wiring replaced by batteries, each of the dozens of water and waste management units throughout the ship are totally self-contained and designed for the 0-g environment. They include showers, toilets, sinks, wet sauna, cooking and laundry units, medical support, fish tanks, and all other fluid systems.

Each unit would have its own recycling, purification system, and reserve water supply. When the supply is low or needs to be replaced, including solid waste, a service bot will service the unit. Waste will be brought to a special recycling system, which would be part of the overall biosphere system, and reintegrated. The servicing of the water units would be accomplished when passengers are not in their cabins or public areas.

Self sustaining biosphere: The science and art of creating and maintaining a structure as a self-sustaining biosphere is advancing at an accelerating rate. This is good news for any future space-based facility or ship, because it must be self-sustaining in order to be economically viable. Our inflatable equilibrium structures must grow their own food, recycle their human and other waste, clean their air, and renew their energy reserves as efficiently as possible. Over the 30-year operational life span of a space facility or spaceship, reducing operating costs by even 10 percent, thanks to self-sustaining measures, will provide a large savings. A biosphere with 95 percent self-sustainability would create a huge cost savings and could be the key to profitability. This will be especially true for large space cruise ships, resorts, and Moon and Mars communities.

Today's ocean cruise ships generate much of their fresh water by distilling salt water. They are required to compact and return all their trash to port and not dump it into the sea. New designs for cruise ships and large ocean yachts are exploring more aggressive recycling systems and, some day, will examine the even more aggressive approach of biospheres.

Air flow compensator (bugs): A vital safety concern in space vehicle design is to provide adequate air flow for living in zero gravity environments. A sleeping person, or one working for a long period in one place, would face a serious health challenge from carbon dioxide buildup. The Space Shuttle and International Space Station deal with this issue by using a system of air ducts throughout the habitable areas. They work, but cause a high level of noise from the air flow and complexity in the interior design. They also add a large amount of mass.

There are no air flow, heating, or cooling ducts in *Destiny*. Instead, mobile air flow compensators, or "bugs" are stationed in volumes and silently follow a person or group of people to maintain the air flows and temperature desired. They will have an extensive sensor system tuned to the specific person/persons in their volumes and their preferences, but they will always keep the air flowing. The bugs will be about the size of basketballs but flattened to a depth of six inches.

Bugs will have the ability to change their color to match any surface to which they attach, being as unobtrusive and silent as possible. They can also free-float if required, for better positioning, and will respond to passengers' voice commands.

This localized and interactive environment control and monitoring system will provide the optimal and safe environment for passengers and crew.

Maneuvering modules (MMs): Like the International Space Station, *Eagle* and *Destiny* will not have dedicated engine volume or its own main engines. (The International Space Station does have thrusters, but relies mainly on the Space Shuttle and Russian *Soyuz* and *Progress* vehicles to reboost and help maintain its orbit.)

My concept of the maneuvering modules follows the same design philosophy as the "less is more," no wiring and no plumbing design of *Destiny*. MMs are totally self-contained hard modules in different sizes and shapes. A typical module will be about two-feet wide, two-feet deep, and three- to four-feet long. They contain several thrusters; fuel; and all necessary support, guidance, and control systems; and are operated via wireless command from the bridge of a ship or facility. Several modules would be attached to *Destiny*'s hull in strategic locations and used in unison to maneuver. When the module is low on fuel, it will be refueled by a specialized service bot. When the module needs major service, or needs to be replaced by a newer model, it will be removed and taken to a service station to be worked on in a shirt-sleeve environment. Some upgrades of systems can be done in place by replacing components. If there is a malfunction and threat of an explosion, the entire module could be jettisoned from the facility or ship hull.

My maneuvering module concept was inspired by studying new cruise ship propulsion systems. The latest system for the large cruise ships consists of propulsion pods attached to the ship hull with a propeller. The pod is electrically powered and can be rotated 180 degrees. This provides excellent maneuverability and is a real advantage in docking ships. The ship can literally move sideways to and from the dock. This system is now in service and is proven highly successful. It ends the use of propeller shafts that added weight to the ship, caused vibration problems, and added significant maintenance problems. This approach is a real breakthrough in ship design.

Nuclear power and fusion: Both technologies provide excellent sources of power for off-world facilities, bases, and ships. They cannot contaminate our Earth's biosphere and their waste products, if any, can be easily managed, stored, or destroyed off world. A small nuclear reactor could be the main power source for a yacht to recharge batteries.

Since I am modeling off-world facilities and spaceships after ocean ships, we can clearly include nuclear-powered submarines in the model. The U.S. and British Navies have excellent safety records in operating nuclear-powered surface ships and submarines since the 1950s. Their experience can be directly applied to the design and operation of reactor off world. Significant research and engineering design and testing have been done for small reactors. As technology and design tools advance, smaller, lighter, and more powerful reactors can be developed. Many of our unmanned solar system probes have been nuclear powered, providing precedent for the safe use of nuclear propulsion in space.

The first generation of space yachts and yacht clubs could be nuclear-powered. The second generation could be fusion-powered, if that kind of system can be made operational. Security for such systems is much easier to provide off world than on Earth, and should a major accident occur, any resulting radiation would not pose a threat to other facilities or ships.

NASA is currently conducting pioneering research into nuclear-powered spaceships and facilities. Those advances can be spun off into the private sector. A whole segment of space advocates are promoting the use of fusion for space power because a key element Helium 3, can be surface-mined on the Moon, which could provide an economic reason to return to the Moon.

There is no reason not to consider nuclear and fusion power for off-world facilities and ships. I am confident these systems will be used in the future, if not by the United States then by other countries or licensed corporations.

Nanotechnology could also change everything in the future. It is the process of engineering materials and machines out of individual molecules — design at the level of atomic dimensions. I have not discussed this subject here because it is such a big topic and the science is still in the early stages.

I could add several more key enabling technologies to this list, and I am sure some of you reading this book may also have excellent suggestions.

Training and Operations

The safety of the passengers and crew must always be the main concern of the yacht and yacht club owners, operators, and Space Guard Service. Space crew and ground staff training, update training, and emergency drills will be key to safe off-world operations.

The uniqueness of the space environment is only part of the space tourist experience. It is the quality of services provided, and the enthusiasm with which the crews and their support personnel greet and interact with passengers that will generate repeat visits.

Earth and orbital-based space academies will provide the core training for command, engineering, operations crews, and ground-based staff. Most of the command and engineering crew members will be retired astronauts or cosmonauts, many with military training, including Space Guard Service personnel. The passenger service personnel will come from the ocean cruise lines, airship cruise lines, and luxury resort industries, including the all-important chief chefs who must constantly provide five-star meals and services in zero gravity.

These specially trained and experienced personnel must be highly compensated for their unique and dangerous work, and should have forms of profit-sharing and equity-building programs.

The immersive space simulation industry will provide many passengers with an Earth-based training of sorts — a glimpse of what real off-world tourism will be like. Some advanced simulation programs could be linked with real orientation programs as cooperative marketing ventures.

Passenger orientation for spaceflights would begin at the Earthport, then continue onboard the orbital yacht club, and finish upon arrival aboard the orbital super yacht. Demonstrations on dealing with zero gravity will be provided, as well as emergency lifepod location/operations, briefings on donning spacesuits, all before the yacht is allowed to undock from the club.

As on Earth, the use of English as the main tourism language and for signage, will be transferred off world, assuming that the United States is the main developer and operator. Remember what I wrote earlier about the danger of assuming the United States will take the lead in the space tourism industry. It is more likely that a new language comprised of slang, phrases, and terminology from a number of spacefaring nationalities will emerge.

Destiny-class yachts can accommodate 20 to 25 passengers, as well as 10 human crew members, plus their Artificial Intelligence Sentinels. Dozens of off-ship telepresence personnel will provide limited-time special services from locations based in orbit and on Earth. This process reduces the number of human crew members required to operate the yachts and club.

Space crew members will be cross-trained, with most mastering two to five positions in addition to their main profession and responsibility. This approach again reduces the number of crew members, while providing a more interesting and productive work experience for the crew.

A typical yacht crew will consist of two command, security, and administrative personnel, and three engineering and operations personnel to manage 100 service bots, conduct physical maintenance, and maintain the biosphere and food production farms. Five crew members will be in direct passenger service, including the ship's doctor and nurse, chief chef, servers, bartender, salon employees, gym staff, store manager, and entertainment staff. Command and engineering crew members will also provide direct passenger services during peak service periods such as dinnertime, arrival, and departure.

Some passengers will want to participate in basic ship operations such as harvesting food from the hydroponics and fish farms, cooking, and providing some of the live entertainment. They will be encouraged and assisted by the crew.

Some passengers may bring their own staff, including personal assistants, butlers, chefs, and security personnel. Some will also bring, or the yacht operators will provide, experts in a wide variety of fields such as astronomy, Earth observation, space sports, and super star entertainment.

Children will be welcomed aboard. They will need to have specialized spacesuits fitted to their smaller stature. Babies under the age of one year might not be allowed in space due to the zero gravity environment. Some pets such as cats may be allowed aboard, with special services to be provided for feeding and waste management.

The key to economical off-world operations will be the extensive use of robots, telepresence, automation, along with AIs and human crew members cross-trained in several disciplines.

Conclusion

Destiny is a great design adventure. Already she has provoked an entirely new perspective and vocabulary for space architecture and operations. She is a platform from which spring new areas of inquiry, because she forces me to ask questions I would have never asked if she did not exist.

And *Destiny* has become a rallying point, as others join my *Destiny* expedition to the future over time. Not all are architects or engineers, but instead represent many professional fields. You are welcome to join the adventure and explore the design frontier with us.

CHAPTER 6
THE SIMULATION FRONTIER

"The tail wagging the dog"
Unknown

Introduction

Simulate: "to have the appearance, effect or form of."

Imagine yourself arriving at the private launch center for the vacation of your lifetime. You check in and drop off your luggage. You change into very comfortable flight coveralls with lots of pockets, then meet your flight director and his or her robot assistant who then guides you and the other nine passengers through a briefing and training session. Learning how to put on a spacesuit can be challenging.

Your group enters the new laser launch capsule. The copilot runs through safety procedures; then the traditional countdown begins, and the power of laser light pushes your spacecraft off the launch pad into low Earth orbit. It is a surprisingly smooth ascent. The views out your viewports and seat monitor are incredible.

Your ascent capsule docks at Orbital Transfer Hub 3 located 200 miles away from Earth. Exiting your capsule into an airlock and then into a "floatway," you encounter your first large viewport view of Earth as she spins by — beautiful and amazing. You are finally in Earth orbit. After checking in by showing your space passport, your robot guide takes your group to an orbital transfer ship that then takes you 200 miles farther away from Earth, to the first orbital yacht club, Eagle. The human-crafted Eagle is a private mini-spaceport, with berths for up to six orbital yachts and all needed services.

Docked already is Ambrosia, a Destiny-class orbital super yacht designed for 20 passengers, 10 crew members, and numerous robotic and telepresence service bots. Ambrosia is a beautiful spaceship composed of three inflated pearl-white spheres connected to each other by a royal blue keel and a light gray hub with gold and silver rim.

Your orbital transfer ship docks with her and you board through an airlock. Your first sensation is the smell of the flowers that are part of her total biosphere system. For the next three days, you and your new yacht mates and crew will yacht around the Earth, enjoying wonderful new experiences guided by the captain, crew, and artificial intelligence sentinel. Ambrosia's sentinel has chosen a female persona and calls herself Jasmine. She loves to sing and has one of the best off-world voices.

After all the passengers have settled into their private cabins, the captain calls all aboard to the yacht's dining, bar, and main lounge for introductions and the famous disembarking ceremony. The views from the lounge are awesome.

After the welcome, the captain goes to the bridge and, with Jasmine's assistance, undocks from Eagle. Ambrosia *is now free to start her fantastic voyage.*

You fly around the huge assembly site of the orbital port, Arthur C. Clarke, *and witness the launch of the latest and largest space cruise ship,* Orbital Ecstasy. Ambrosia *will join other ships escorting* Ecstasy *on her maiden orbit. Next are visits to space sports facilities; then you participate in a thrilling orbital regatta race.*

Ambrosia *also offers you the choice to take a space walk ("float"), participate in scientific experiments, harvest or catch your own dinner from the hydroponics garden and fish farm, have a massage, try out a new hair style or body-paint, exercise or use the sauna.*

Attend a wedding; then party all night long with live entertainment and four-star food, drink, and service. Call your family and friends back on Earth to let them know what a great time you are having and that you are flying over them as you speak. Have videos and photos taken of your orbital yachting experience to show them once you return home.

Have private time in your cabin and enjoy crystal clear views of the Earth, Moon, and far space. Relax and stare out at the Earth moving by while listening to your favorite music. In just over 90 minutes, you will enjoy a brilliant sunrise and a sunset, highlighting the thin blue band of atmosphere protecting all life on Earth from the cold of space. Orbital yachting and cruising is the ultimate eco-tour, offering the opportunity to see most of our beautiful planet. Many people return from orbit with a new or renewed dedication to protect and preserve our only spaceship Earth.

After returning to Eagle, *and saying your goodbyes to* Ambrosia*'s crew and* Jasmine, *you return to an orbital transfer hub and then to a laser capsule that takes you through your exciting, fiery return flight to Earth. Once back home, you can choose to participate in a few hours of relaxed debriefing and farewells with your yacht mates or just head home after visiting the gift shop. You will receive an excellent discount if you book your next off-world experience before leaving. During your three days on orbit, you circled the Earth 48 times, traveling almost 1.5 million miles, equal to traveling to the Moon and back to Earth three times.*

That is the fantasy. However, in reality, you never left the launch center building. You were at a Simcenter, where the entire orbital yachting experience just described was an elaborate precreation of orbital yachting, decades in the future. It was a simulation designed to create a totally immersive and facilitated themed experience.

You and the other passengers were "Simnauts," participating in an elaborate interactive play. At the same time you were contributing to the designers' scientific knowledge of human factors-design and space architecture, while assisting those who are researching and planning the design of real orbital tourism ships and facilities in the future.

The Simcenter immersion process begins with reading and virtual sim computer play, well before arriving at the physical sim. There you are totally separated from the rest of the world for at least two full days. Waking up in that environment is a highly significant part of the immersion experience, as is wearing the clothing of that theme, eating the food, listening to the music, using the language and devices, and interacting with the crew or staff and other players. In a fully immersive experience, you are a participant in an interactive play, living in an authentic world/stage. Every sound, view, taste, smell, activity, vibration, color, and design is carefully staged to enhance the believability of the experience.

Sophisticated sims with beautiful spaceship interiors and fantastic group activities are now under design and development. They are on the leading edge of the next generation of high-end vacation and entertainment development — the "SimExperience" industry.

SimExperience is a word I created in the mid-1990s to describe the merger of simulated immersive experiences with character-driven, high-quality overnight experiences. I created the word "Simnaut" to describe a passenger who is participating in these immersive simulations. "Simcrew" describes the crew inside the simulated environment and the ground crew at the "Simcenter" back on Earth.

In a few years, you will be able to take simulated orbital yachting vacations. By the end of this decade, you can join Simnauts at larger "Simports" to take space cruise ship and lunar cruise ship vacations. Take your whole family for a week-long vacation at a Moon or Mars resort and spa. Join the heroic crew of a Space Guard Service space cutter saving stranded space tourists and defending the Earth from meteors or comets, anticipate first contact with aliens, or even become a space pirate. The possibilities are endless for the emerging immersive simulation industry.

I am highly confident a SimExperience industry will grow over the next few decades to become a significant segment of the multibillion-dollar adventure/expedition vacation industry. This confidence is bolstered by the recognition of respected analysts and forecasters in the tourism and entertainment industries who have identified the experience economy as the next mega trend in American society.

In essence, people want to participate in meaningful vacation and entertainment experiences, ones through which they can master new skills, learn new things and interact with experts and others of similar interests. They want experiences so intense, challenging, and rewarding that they forget their busy and stressful lives for a few days of exploring unique and different social and physical environments.

Baby boomers and senior citizens are willing to pay top dollar for new kinds of experiences. Adventure vacations, themed resorts, and expedition cruise ships are growing in popularity, as are the large variety of successful space camps for both kids and adults. Dude ranches, amateur scientific expeditions, Greek and Roman villa spas, renaissance pleasure faires, rain forest tours, African photo safaris, Civil War reenactments, and many other examples validate the growing trend of participatory experience vacations.

The space-themed simulations are, by their very nature, small, controlled environments and, therefore, ideal as a main theme for the SimExperience industry. Building a successful immersive space simulation industry is the ideal strategy to building market awareness and eventual demand for the real thing. The first few dozen real space tourists who visit Earth orbit (especially if they are major celebrities) will generate world-wide media attention, thus promoting the space tourism-themed simulation industry.

Clearly a highly symbiotic and beneficial relationship exists between the real and simulated space tourism industries.

Space Simulations Today

At least a half a billion people have taken trips into space already. Simulated trips, that is.

Since the early 1960s, the world has been learning about space travel, exploration, science, and now space tourism. By watching television news programs, movies like *Apollo 13* or *Mission To Mars*, visiting space museums and NASA centers, or reading, the general public's level of space awareness is high. You could list between 20 and 40 space terms right now without having to think too hard. Try it. Before reading on, make a list and test your level of space awareness.

Did your list include terms such as countdown, launch pad, liftoff, orbit, astronaut, cosmonaut, spacewalk, spacesuit, space shuttle, zero gravity, rendezvous and docking, airlock, *Apollo* program, lunar landing, Moon rock, lunar rover, reentry, space station, or *Hubble* telescope? Maybe your list included famous names like John Glenn; Neil Armstrong; Kennedy Space Center; *Challenger*; *Columbia*; *Mir*; and the Mars rovers *Sojourner*, *Spirit* and *Opportunity*. Good job! My point is that we have all observed the world's space programs for most of our lives. Space is part of our culture. When we think about it, we know more about it than we realize.

This knowledge and, to some extent, expectations of space are great advantages for the designers and developers of space simulations. We already have an audience familiar with our theme, the general history of space, many of its common terms, and what it should be like when we go there. In the simulation, small groups of people enter entirely controlled environments that take them off world for a mission and return them to the same place from where they started. Experience meets expectations. When passengers look out viewports, they will see views of space and Earth with black backgrounds that facilitate visual effects. Simulated orbital cruise ships, Moon and Mars bases, and resorts crews will encourage passengers and visitors to put on spacesuits and take spacewalks/floats, or explore the surface areas that are also highly controlled themed environments, with expert guides facilitating the experience. For many, these space floats and surface walks will fulfill life-long dreams and be their peak experiences of the sim.

The unique aspect of off-world travel is experiencing zero, lunar, or Mars gravity. The one drawback of space-themed simulations is that we cannot create zero or low-gravity during the entire simulated environment. We can only simulate zero or low-gravity in special areas of the sim by using special equipment.

Fortunately, hundreds of millions of dollars, and large amounts of engineering talent and time and testing over the past four decades have created numerous methods and devices to simulate off-world gravity. Many of these methods and devices have been simplified and used to entertain and train the public.

The movie industry has pioneered several methods of defying gravity, most by suspending actors from thin wires. The special effects team on the HBO miniseries *From the Earth to the Moon* developed a unique system of using 12, four-foot diameter helium-filled balloons tied by one wire to the backpack of an *Apollo*-era spacesuit, which allowed the actor/astronaut to bounce on the lunar sets. They accurately simulated the one-sixth gravity of the Moon, creating some impressive film sequences.

Zero and low-gravity experiences will be one of the highlights of any space simulation. For the areas of the sim not able to simulate 0-g effects, a variety of techniques will be used to remind passengers of their special environment. Lids must always be placed on cups to keep liquid from escaping and floating away. Velcro is on every device and surface to provide anchoring points. Simnaut shoes will have soft soles, and a "floor" will be covered in spongy matting to provide the impression of slight bounces with each step. Lights and signs will be positioned in ways to neutralize normal up and down orientation. Even the colors and overall interior design will be selected in order to simulate a more three-dimensional environment.

Walt Disney designed and built one of the first space simulators for the public. Opened in Disneyland's Tomorrowland in the late 1960s, it was called the *Trip to the Moon* attraction. In the 1980s, it was renamed *Trip to Mars*. The attraction featured a mission control (queuing) area leading to a rocket vehicle consisting of a deep round chamber with circles of seats and view screens on the floor and the ceiling. It seated about 200 people. When the rocket took off, the seat bottoms would compress, giving the very slight illusion of g-force. I was 10 years old when I first entered that magical chamber and experienced my first trip to space. It changed my life, intensified my interest in space and science, and jump-started my personal quest to go someday.

In the late 1980s, the first *Star Tours* simulation ride opened on the West Coast in Disney's Tomorrowland. It is themed around the popular *Star Wars* movies created by George Lucas. The attraction has an elaborate queuing area, which represents a spaceport passenger terminal. Large view screens show exotic planets and locations open for tours. Loudspeakers announce star tour vehicle arrivals and departures.

There are four shuttle simulators, each resting on platforms with powerful hydraulic legs that rapidly move up, down, and sideways. The passengers (space tourists) never see the legs. Each simulator holds 40 tourists in theater-style seating, giving everyone excellent views of the forward view screen. Everyone must wear seat belts or the ride will not start. The visuals are coordinated with the movement of the platform, giving a wild 4.5 minute ride through the *Star Wars* universe, hosted by the robots R2-D2 and C3PO. The visitors exit the simulator into a long ramp leading into a large gift shop. The attraction can process 1,500 people per hour. Today there are *Star Tours* attractions at both Disney parks.

The first use of this motion platform and visual project technology for the general public and entertainment was at the *Tour of the Universe* simulation ride at the CN Tower in Toronto, Canada, which opened in 1985. It was a successful proof of concept on which the Disney people later modeled their *Star Tours* attraction.

In October of 2003, Disney opened *Mission: Space* at their Epcot Center at Walt Disney World in Florida. This $120 million pavilion was sponsored by Hewlett-Packard. It takes visitors on a simulated ride to Mars and returns them to Earth. NASA and retired astronauts were consultants on the design of the attraction. The Disney company expects between three and four million people to experience it each year. It has proven highly popular, so it may be reproduced at the other Disney parks around the world.

The first park to be constructed solely with a space theme is called Space World. It opened in April 1991 in Kita Kyushu, an industrial city in the southern part of Japan's main island. The park is owned and operated by Nippon Steel in joint venture with Mitsubishi Corporation. In the center of Space World is a full-scale, highly detailed, vertical

Space Shuttle stack with external fuel tank, solid rocket boosters, and shuttle orbiter. The park attracts close to one million visitors per year to its two-dozen space attractions, IMAX and other theaters, amusement rides, and shows. There is also a space camp, space museum and facilities for educational programs.

The Space World park was the result of the joint venture between a space simulation design and development company I founded in 1982, called Space Resort Enterprises, and the Mitsubishi Corporation. The joint venture, formed in 1988, was called Spaceport Systems Corporation, whose purpose was to build the Space Resort project I had created and other space attractions around the world. The Space Resort was a 220-suite luxury resort located under a launch pad with a full-scale Space Shuttle stack and gantry tower on top. Passengers would take a simulated ride into orbit in the shuttle, and stay immersed inside the Space Resort between three and five days, enjoying all the activities of a cruise ship in Earth orbit.

Another project that resulted from this joint venture was a full-scale Space Shuttle experience ride at Six Flags Great America theme park in Chicago, built in 1995. Based on a concept created in the early 1990s by long-time colleagues Jason Klassi and the late Charlie Carr, the simulation invites visitors to enter a hangar building located next to the shuttle. While visitors believe they are entering a passenger module inside the shuttle cargo bay, they are actually entering a theater resting on a motion-based platform. Next come liftoff and a wild trip around the Moon. More than 1,000 people every hour take this exciting space simulation experience. The architects for this project were the Cuningham Group with Rick Solberg, AIA as the lead designer. They were also the same team who designed *Star Trek — The Experience* noted below.

Of course one of the granddaddies of the simulation experience is the *Star Trek Adventure* attraction, which was opened by Universal Studios in Los Angeles in the mid 1980s. It was one of the first attractions to involve audience members in the actual show. While waiting in line, about 30 members of the audience were selected to be cast members on stage. Dressed in Star Fleet or Klingon uniforms, or made up and costumed as aliens, they were directed by the show director to perform short scenes on three different stages located next to each other.

Most of the time this impromptu cast would flub their short lines or be in the wrong place and miss cues, which delighted the live audience. Suspended above the set were large screens displaying scenes from some of the *Star Trek* movies. A then revolutionary computer-driven editing system would be used to merge footage of the recruited cast members with real movie scenes and music and, by the end of the 30-minute show, produce a five-minute videotape to screen to the audience. About 100 tapes were sold at each of the day's four shows at $35 each!

This particular *Star Trek* themed show is no longer being produced, having been replaced by other themes and far more sophisticated editing systems. But these kinds of audience participation shows remain the highest-rated at the theme parks.

One of those new attractions is *Star Trek: The Experience*, which went to the next level of immersive experience. This 70,000-square-foot, $75 million attraction opened in January 1998 at the Hilton Hotel and Casino in Las Vegas, quickly becoming a nirvana for Trekkers and an all-around fun experience for the general public. Once visitors enter this attraction, they are immersed in the *Star Trek* universe of the 24th century. The entire attraction is populated with dozens of walk-around characters in Federation, Klingon, Vulcan, and, most recently, Borg costumes. These actors, who are a highlight of the entire experience, are well-trained in the Trek universe, attitudes, and languages, and work hard to interact with the visitors.

The attraction's *Museum of the Future* showcases props, art, and designs from all the *Trek* television series and movies. The museum is used as a queuing area leading to the beaming station, which "beams" small groups of visitors to the bridge of the next generation Star Ship *Enterprise*. The visitors become crew members and, after leaving the bridge, take a shuttlecraft on a wild ride, where they are attacked by rogue Klingons and barely escape to the famous *Deep Space Nine* spaceport. After leaving the craft, they tour the three-story-high promenade leading to Quark's bar and restaurant, which leads to a large gift area selling more than 1,000 *Trek* items and gifts.

Star Trek: The Experience is well designed and stays faithful to the *Star Trek* universe. It is actually a spin-off of a concept I originated and designed in 1989 called *Star Base One*, the first *Star Trek* theme park. I knew Gene Roddenberry, the creator of *Star Trek*, and had suggested we develop a "Trek Park" at the height of *Trek*'s popularity. Gene was an advisor on the project until his death in 1991. For a variety of reasons, we did not build our park, and I was not involved with the attraction in Las Vegas, but I know the designers and architects and feel they did a great job. My long time friend and Paramount Studios, chief production designer for *Star Trek* since the mid 1980s, Herman Zimmerman, spearheaded the design for this project from start to finish.

Rounding out our gallery of Earth-based space simulation experiences are:

LunaCorp (*www.lunacorp.com*) is building an electronic bridge between real space exploration and space simulation. Founded by David Gump, LunaCorp was later joined by Jim Dunstan. Together they have pioneered the concept of tele-exploration and tele-tourism. They plan to privately finance a mission to the Moon, landing a rover near one of the historic *Apollo* landing sites to explore the site and surrounding areas. For a fee, the general public will be able to drive the rover while sitting in a driver's chamber resting on a small motion platform that mimics the rover's motions on the Moon. These chambers will be located at science centers.

In the late 1990s, LunaCorp successfully tested their concept, software, and technology by completing a 220 km exploration of the Valley of the Moon in the Atacama Desert in Chile. Working with the Robotic Institute at Carnegie Mellon University, their *Nomad* rover was directed and driven by people thousands of miles away. It was an excellent proof of concept for their private lunar exploration venture.

Space Camp (*www.spacecamp.com*) in Huntsville, Alabama, has pioneered the space mission training experience for both kids and adults. More than 10,000 kids between the ages of 12 and 17 travel to the camp each summer for five days of training and simulated space missions inside full-scale mockups of the Space Shuttle and International Space Station. The space campers are divided into shuttle crews and mission control crews who train for three days and then perform realistic space missions. There are several space camp operations around the world.

The Challenger Learning Centers (*www.challenger.org*), founded by the families of the astronauts lost with the Space Shuttle *Challenger* in 1986, has built more than 50 Challenger Centers around the United States and have 20 more in development. Each 2,500-square foot Center consists of a briefing area, a mission control area, and a space station simulation area. Middle school classes are given full-day experiences of being both the mission control operators and the space station crews. They learn science, mission planning, teamwork, and general space history. Most important, they glimpse the potential of further study or a career in space science, engineering, or math. The Challenger Learning Centers also offer fun simulation programs for adults as part of corporate team-building efforts.

There are other space simulation facilities around the world at space museums and science and technology centers, with a growing number of larger ones in development. The common goal of all these space simulation providers is to put the audience or participants into space and to entertain them with a space experience on Earth. What is wonderful is that, so far, many of them provide surprisingly realistic space experiences.

The Experience Age

There are clearly definable trends in the United States tourism and entertainment industries that confirm the public's strong desire to participate physically in its leisure, vacation, and entertainment experiences. No longer satisfied to passively watch television, sports events, and movies, some people want to be directly involved, to influence the outcome of situations, to have a voice, and to play a role. To feel what it is like to be in the experience, not just an outside observer.

We have entered what I call the "Experience Age." This is an era built upon hundreds of years of tourism and entertainment development in Europe and the United States — an age oriented toward high-end markets, specialized theming, small groups, with a story-oriented and participatory experience.

These experiences are enabled by a new comprehensive design approach that integrates a themed physical environment with a themed social environment and character-driven interactive stories. The goal is to create meaningful, challenging, and joyful experiences that will be fondly remembered (while making a profit for the developers and operators in the process).

In many ways, Walt Disney can be considered the founding father of the Experience Age. He and his talented team came from the movie business, so they were trained in both the art and science of storytelling. With his ultimate dream of a theme park beyond any amusement park ever envisioned to that time, Disney embarked on producing a television show that would be used to finance the project. In July 1955, Disneyland opened its gates to the public and literally billions of people and billions of dollars have flowed into this and other parks around the world ever since. The Magic Kingdom was where his stories, characters and places came to life, interacting with audiences and visitors from around the world.

Disney invented the theme park and an entirely new way for the family to enjoy traveling together to different places and times — Frontierland in the past and

Tomorrowland in the future are just two examples. Disney established a high level of thematic integrity for the places, characters, costumes, performances, music, and staging that set benchmarks for everyone in the theme park and location-based entertainment industry.

One breakthrough of the Disney parks was the use of stories and characters from their successful animated and live action movies and television shows. Developers could test stories and characters first on film, then, if they proved popular, build a ride or show using the same themes. Visitors to the attractions would already know the story and characters, and how to interact with them. They would feel immediately comfortable, especially children who had seen the movies and read the books. All Disney visitor surveys reveal that the favorite part of visiting Disneyland for kids is meeting Mickey and Minnie Mouse and the other walk-around characters, playing and having their pictures taken with them.

The Disney team pioneered the development of new technologies and techniques to tell stories first for their animated films, television, then live action films, and finally their theme parks and themed resorts. They also invented what is called "the dark ride," such as *Pirates of the Caribbean* and *The Haunted Mansion*, which took an audience on a slow ride through a themed environment while telling a thrilling story.

While the Disney people of the past and of today do not use the term simulation in their theme parks and Epcot Center, the Disney parks are, in my opinion, simulation centers on a grand scale.

Epcot Center's *Living Seas* pavilion is a perfect example of a simulated themed environment and is currently one of the best in the world. Sponsored by United Technologies, this impressive $100 million pavilion opened in 1986. The series of entry experiences is designed to place the visitor in the frame of mind where they are going to explore a beautiful undersea world. Visitors enter hydrolaters, which look like large elevators with glass sides. They descend, while noise bubbles rush up the sides and the floor vibrates, down to SeaBase *Alpha*. Actually, the hydrolater does not move at all. It is an illusion, just like the supposed room that "stretches" down into the catacombs of *The Haunted Mansion*. When the hydrolater "stops," doors opposite the original entry doors open up and the explorers are welcomed aboard Sea Base *Alpha*.

This futuristic setting is a two-story, beautifully designed and detailed seabase, located in the middle of a 200-foot diameter, 24-foot-deep, 5.7 million gallon saltwater tank, that is home to thousands of fish, dolphins, manta rays, sharks, and a multitude of other sea creatures, along with an artificial coral reef teeming with a wide variety of plant life waving in the slowly circulating water. Floor to ceiling viewports provide stunning views of the sea world below and above the visitors. It is a total immersion experience generating many "oohs" and "aahs" from the visitors.

Sea Base *Alpha* is full of displays of sea exploration vehicles, some real and some imaginary, dive suits, and ocean robotic rovers. Uniformed crew members are eager to explain the displays and the wonder of ocean exploration. Every 15 minutes, loud speakers announce that an Aquanaut will be going outside to work with some of the sea life. He or she strides through the SeaBase in full diving gear, waving to the visitors, and then enters a glass water lock and enters the sea.

Visitors tour *Alpha* at their own pace, typically spending about 30 to 40 minutes in the pavilion before exiting. There is also a restaurant accessible from outside the pavilion with great views of the artificial sea and the Sea Base.

A large number of marine life scientists, oceanographers, ocean engineers, and other specialists from dozens of universities, institutes, and companies were engaged to help the Disney team design and operate the *Living Seas* pavilion. Actual scientific research is conducted there, in the midst of an entertainment park.

The following are a few examples of other themed experience environments and activities that qualify as entries in the Experience Age:

Renaissance Pleasure Faires were first staged in California in the 1960s, and always in an outdoor park or pasture. They are temporary villages of 50 to 90 tents, booths, colorful wagons, jousting tournament areas, stables, and stages that occupy three to five acres. Faires run about eight weeks, then move around a state or country during the summer season. For some of the actors and artists, it is their full-time occupation and lifestyle.

"All the Faire's a stage and you a merry player ... Come join the pleasure!" shouts the headline on a Faire flyer. This Faire's theme is centered around Queen Elizabeth's reign from 1558 to 1603 in England, featuring hundreds of actors in period costume speaking Elizabethan English, with demonstrations of culture, music, dance, food, games, contests, Shakespearean events, petting zoos, parades of the Royal Court, and a wide variety of artists creating period products. Nobles, peasants, minstrels, milk maids, huntsmen, jesters, and fools are some of the characters who populate the Faire. The Faire creates a living history for the public to enjoy where they can study the era, learn the language, make their own costumes and accessories, and directly participate. Role playing is the key part of the Faire experience both for the participant and the public.

Dinner theaters are another example of making something as normal as having dinner a special and memorable experience. Not at all a new idea, feasts and royal dinners have been around for hundreds of years in England and France. Elaborate dinners were sponsored by royalty as major social and political events.

Hawaiian luaus are a simulated experience. Today, one can attend a recreation of a Hawaiian luau performed outdoors on a Hawaiian beach with palm trees and the sound of crashing waves. I can still remember in detail the first one I attended. The flower smell of the lei around my neck, the smell of the roasting pigs in open fire pits, the brilliant flames from fire posts lighting the area, fire dancers, the beautiful Hawaiian women dancing with grass skirts, the rhythmic beat of the drums, and the unique taste of the food. I found the entire experience to be primal, so well staged and performed I totally connected to it on a personal level.

Medieval feasts invite guests to "Come with us on a magical journey to the Middle Ages and return to a time when valiant knights battled to become the personal champion to the king. Enjoy the wonders of our castle and satisfy your appetite with a feast hearty enough for the richest of kings. Come back to Medieval Times!" This quote is from the promotional materials of the Medieval Times dinner and tournament show, a successful themed restaurant/theater business. An audience of more than 1,000 sits in an arena with dozens of actors and horses performing on a dirt floor at least 100-feet wide and 200-feet long. The tournament has real jousting and sword fights, with the winners receiving inspiring praises from the king and queen.

In this experience, the food, while excellent, is secondary to the show. The pageantry, music, blowing trumpets, knights in armor, and wenches dressed in medieval costumes, all speaking the language of the era, and all contributing to a fun and exciting evening.

Deluxe movie theaters are popping up in major urban centers around the country. There are several of these type of theaters in Los Angeles, seating an exclusive audience of only 25 in soft ergonomic seats, with state-of-the-art projection and sound systems. The ticket price is at least twice the cost of a normal ticket, however price is not an issue. These theaters serve hot *hors d'oeuvres* and wine, and offer ushers wearing white gloves. Some have live music before and after the movie. Others will even pick up audience members in a limousine and return them after the show for an extra fee. The theaters promote the exclusive nature of the venue and the total movie-going experience.

Mega-bookstores such as Borders and Barnes & Noble offer perfect retail examples of the Experience Age. By providing comfortable seating areas, work tables, good lighting, cafes, and an attitude that the customer is welcome to stay as long as they like, the stores have made shopping for a book or magazine a friendly experience. In fact, part of this book was written at Barnes & Noble in Santa Monica because it was such a comfortable place to work. Quiet, but not isolated, as lots of other readers shared the time reading and writing.

The Experience Economy

A key player in the emerging Experience Age is the "Experience Economy," a term coined by Joseph Pine and James Gilmore, authors of the book *The Experience Economy: Work Is Theater and Every Business a Stage* (Harvard Business School Press, 1999). I highly recommend this book if you become interested in the emerging Experience Age and SimExperience industry. Pine and Gilmore recognized that corporations, retailers, and educators were far more successful in productivity, sales, and learning when the participants engaged in the experience rather than just observed it. By connecting people in a participatory manner, they focus on the experience, generating more creative solutions in business situations, buying more products and services in retail situations, and attaining better comprehension in learning situations. By actively participating, they are less concerned with the time or money spent and more concerned with having a meaningful, personalized experience. The author's conclusion was that those who provide an interesting, enjoyable, and meaningful experience will be more successful than those who do not. Their company is called Strategic Horizons (*www.strategichorizons.com*).

The Experience Age and Experience Economy establish a strong foundation for the SimExperience industry. Immersive overnight experiences, where participants study and prepare for the experience and are willing to pay top dollar for truly meaningful participatory experiences, are clearly the next generation in vacation and entertainment business development.

Immersive Experiences

The next generation of simulated experiences will be entirely immersive. The Simnauts will be highly motivated and well prepared to assume roles in interactive plays performed in totally controlled environments staged over two to seven days. They will live in the SimExperience, taking responsibility for the quality of the immersive experience facilitated by a highly skilled sim crew and support staff.

Immersive simulations are certainly not limited to the space theme. Any theme that can be staged in an entirely controlled environment for multiple days will work. Thematic examples of other immersive experiences include:

- Submarine adventure
- Underwater science base
- Simulated cave spelunking
- Egyptian or Mayan tomb exploration
- Haunted houses
- Sea colony
- Treasure hunting
- Luxurious airship cruises
- Knights of the Round Table
- Spy thrillers

This wide thematic range provides the emerging SimExperience industry with excellent growth potential. It also enhances international growth potential of cultural themed sim ideas including a Samurai training camp, an African hunting ritual, a South American Inca temple ritual, and so on.

The simulated orbital yachting scenario described at the beginning of this chapter best illustrates the space-themed version of an immersive experience. What elevates and intensifies the immersive experience over a simulated experience are these three requirements:

1. The overall story of the SimExperience must be believable. The characters, context, physical environment, language, and sim culture must make logical sense in order for the Simnauts to make emotional connections to the new world they have entered.

2. The sim must last at least 48 hours so the Simnauts will wake up immersed in the Sim. Sleeping in the sim is critical to creating both the emotional and physical connection to the characters, culture, and physical place. Three to five days will provide the most intense and, therefore, meaningful experiences.

3. The Simnauts, Simcrew, and any visitors to the sim must wear the clothing or uniform of the sim theme. Wearing themed costumes encourages all the participants to more deeply engage in the culture, activities, mission, or quest of the sim. Accessories such as tools, backpacks, communication devices, jewelry, war paint, hair styles, and perfumes contribute to the realism of the sim.

Other important aspects, but not requirements, include:

Preparation: The amount of preparation the Simnauts have before arriving at the simulation facility or location will affect the quality of the overall experience. Some enthusiasts will have tremendous knowledge and even real experience in the theme of the sim. Some will have little or no knowledge of the theme. General briefings will be conducted by the staff so all participants have a basic understanding of the "world" they will be entering.

An important part of the briefings will be an explanation of what is expected from the Simnauts in order to reduce anxiety and enhance the pleasure of their immersive experience. In many sims, the participants will have chosen the level of participation they want. If it is a first time SimExperience, some will prefer to observe more than participate. Others may want to be the star of the sim in which the outcome of the mission, quest, or expedition depends on their leadership and decisions.

Cost: Simulation facilities will be expensive to design, build, and operate. Many will be ultra-small luxury resorts and health spas for 10 to 30 guests. Many will offer elaborate home bases and destinations. For example, a Mars Expedition Base for 16 explorers and Simcrew could be located within a 20-acre Mars-scape with realistic enclosed rovers taking people from Mars caves to other locations within the simulated environment.

Repeat visitation: Each sim experience will be unique because the people involved will each react differently to the simulation and to the other Simnauts. This will enhance repeat visitation because, while the simulation conditions remain the same, the people participating in the sim do not. A person may want to repeat a simulation several times in order to get the most out of the experience and to advance in the ranks of a "Simnaut Corps." New kinds of sims will be promoted through websites, chat rooms, newsletters, and eventually magazines and television shows. Some Simnauts may even want to return as Simcrew members.

The following examples of current immersive experiences fulfill the three requirements of: 1) believable story, 2) multiple-day immersion, and 3) wearing the specialized clothing or uniforms of the theme. Some also encourage or even require participants to prepare for the experience with reading and discussions about the theme or location. All illustrate the growing market demand for high-quality immersive simulated or staged experiences.

Space camps: (described earlier in this chapter)

Wagon train recreations: In 1851, a group of Mormons left Salt Lake City for an 800-mile, two month trek west to San Bernardino, California, to establish an outpost. During September and October of 2001 — 150 years later — a dozen families followed their trek in horse-drawn covered wagons, dressed in period costumes, cooking over open fires, fixing their wagons, and enjoying themselves, most of the time.

Other groups recreate the Oregon Trail wagon trains. Some extreme groups retrace the Lewis and Clark expedition. There are small cattle and horse-herding adventures, and numerous dude ranches where men can be cowboys and a growing number of women can be cowgirls. Some of the groups participating in these very American recreations are from Japan, Germany, and other countries who are fascinated with the frontier period of American history.

Civil War reenactments: Each year, thousands of Civil War enthusiasts gather to recreate historic battles and engagements. Some are staged on the actual battlefields. These recreators have well-organized regiments and platoons. They bring their horses, wagons, cannons, tents, period musical instruments, food, drinks, and everything else to accurately recreate living history. Much research is done to make sure the participants, props, battle maneuvers, time of day, and duration of battles are reenacted accurately.

Most of the participants make their own uniforms, including hand-carved buttons and belt buckles. Some bring real rifles and pistols that were actually used in the Civil War. Some groups train like real militias in their own communities. They march in review and live in tents of the period for several days. Some of the wives and girlfriends attend, also wearing the clothing of the period.

Many of these groups have appeared in Civil War movies and documentaries fielding thousands of uniformed participants. These fun, immersive reenactments have played an important role in preserving a critical period in American history.

There are many other examples of a wide variety of people taking the time and spending substantial amounts of money to immerse themselves in unique places and social settings, while cutting themselves off from their normal lives. While not a new trend, in the past it was usually only the wealthy who could afford to immerse themselves in exotic and unique situations. The difference today is that the average person and family can immerse themselves in an ever-widening amount of thematic, cultural, and physical settings.

The evolution of the immersive experience is clearly toward story-oriented themes where participants can inhabit roles and have direct responsibility for making the experience work. This is an exciting trend because it opens the door to much more intense and fulfilling immersive experiences, and to experimentation in new forms of art and entertainment. However, one must always remember that the key to success will be to have a good story to experience.

The Power of Storytelling

Humans love to tell and hear stories. It is a key factor that separated us from animals and allowed us to become the dominant species of this planet. Before writing was invented, oral storytelling was the only way to pass along knowledge and customs from one generation to another. In some ways, stories are time machines. They can recall great deeds or tragedies from the past or foretell great adventures in the future.

For countless generations, people have gathered around a camp or cave fire for safety, community, and to tell stories, passing myth, legend, and knowledge from generation to generation. They would tell of heroes battling fearsome monsters or traveling beyond the horizon, deeper into caves, or down rivers and across oceans on bold quests.

Anthropologists now believe some early cave paintings were part of elaborate storytelling ceremonies and initiations. Flickering fires made the painted animals appear to be running across the cave wall — perhaps the very first use of special effects!

Immersive simulations are just elaborate stages for stories. The participant interactions and, most important, the story they are acting out, are what will bring the simulation to life. Sim stories give direction, purpose, pacing, duration, and meaning to the SimExperience. Stories are the foundations from which the experiences are built. They must have a clear beginning, middle, and end to provide Simnauts with a real sense of accomplishment and closure. The main story of the SimExperience will establish the mission, quest, or adventure and purpose that bring the participants together.

There is a high degree of failure in all exploration. The risk of failure or great reward and recognition give exploration its special allure and excitement. An example is the heroic story of the failed *Apollo 13* Moon mission. Crippled by an explosion on the way to the Moon with little chance of survival, the crew members, ground control, and all of NASA struggled to get them home. With real courage, determination, training, and wild innovation, the crew just barely got home. Much of the world watched this fight for survival and rejoiced over their safe return.

The story behind the rescue of *Apollo 13* has become an important legend at NASA. In the first briefing at Mission Control immediately after the explosion when things looked hopeless, Flight Director Gene Kranz said, "Failure is not an option." Years later, this phrase became the title of his excellent book, and the unofficial motto of NASA and the private space entrepreneurial community. It was that heroic attitude and the drama of the story that attracted the support of actor/director Tom Hanks, who starred as mission commander Jim Lovell in the highly successful movie *Apollo 13*, directed by Ron Howard.

The quest for space tourism is the story of our time, just as the story of western expansion was the story of that era, and as the journeys of exploration and settlement of America were the stories of their own eras. Fortunately, our story of exploring and expanding beyond Earth is a wonderful and limitless story. It is also a story of the human species and not just of a few individuals or countries. We will be living and telling it for the next 100 years, and highlights will be remembered far beyond our time.

Joseph Campbell is recognized and honored as a central pioneer in the study of myth and mythology from around the world. Like Albert Einstein's pursuit of a unified field theory for physics, Campbell spent 50 years reading, researching, teaching, and traveling the world in pursuit of the common psychological roots of all myth, science, religion, and art. His discussions with some of the greatest thinkers of the 20th century, and detailed study of ancient texts such as *Gilgamesh*, the *Tibetan Book of the Dead*, the *Egyptian Mysteries*, the *Iliad*, and the *Odyssey*, allowed him to conclude that they were all saying the same thing — that there is a system of archetypal story themes that have warned, encouraged, and inspired humankind for thousands of years. Campbell called this system "one grandiose song."

Some of today's storytellers in all media have been influenced by Campbell's books, lectures, films, and interviews about his research. George Lucas, the creator of the *Star Wars* movies and, with Steven Spielberg, the *Indiana Jones* movies, credits Campbell's scholarship and books with giving him the focus he needed to complete the master *Star Wars* saga. Campbell's books include *The Hero with a Thousand Faces*, *The Inner Reaches of Outer Space*, *The Mythic Image*, and *The Power of Myth* with Bill Moyer. The best book about Campbell and his work is titled *The Hero's Journey* by Phil Cousineau. Visit the Joseph Campbell Foundation at *www.jcf.org*.

The reason I mention Campbell and the significance of myth in storytelling is the need for immersive simulation activities to be based on universal quest and adventure oriented stories, so that participants have goals to work together toward achieving. The stories must be as cross-cultural as possible so they are as internationally appealing as possible, and thus provide the SimExperience industry with a wide audience base.

The very nature of space exploration and off-world tourism crosses borders and cultures. So we have an opportunity to create new myths and legends of our time, with our own voices and new stories based on fundamental human nature, desire, and the nobility of going off world.

Sim Academies

The SimExperience industry will eventually need several Sim Academies to provide thousands of specially trained and motivated employees. The Disney corporation can be a model for the Academies. Walt Disney realized his company would need thousands of well-trained and motivated staff members for his theme parks. He considered the staff members who had direct contact with the visitors to be characters within the overall theme park setting and story, so he founded the Disney University, which orients new staff members to the Disney attitude of "always smiling, always cheerful, and always helpful." The Disney attitude is very important to the expectations and satisfaction of the visitors. A disrespectful or out-of-character staff member can ruin the entire visit for the kids and, therefore, the parents.

For the immersive, multiday simulations, the attitude and performance of the Simcrews and staffs will be even more important due to the close interaction with the Simnauts. Simcrews and staffs must be in the proper frame of mind and attitude for the specific themes of the sim. They will need to function as a real team to properly and safely operate the sims. The Simnauts will need to see and feel that the Simcrews and staffs work well together. This will encourage the Simnaut to symbolically join the crew and make the whole experience successful for all. For the space-themed sims, creating a mission name and mission patch like real astronaut crews do will contribute to team building and successful sims.

Combining training methods of the Disney University with those of the U.S. Coast Guard and the proven training programs of the cruise lines could form a solid foundation for the Sim Academy.

There are a growing number of prestigious universities and institutes already offering classes in space tourism development and operations. They could make a real contribution to establishing a high-quality Sim Academy. Below are a few examples of allied University programs.

The International Space University is currently the closest program to a real space academy. They have plans to expand and are interested in including space simulations as a study subject. They have conducted numerous space tourism studies and projects.

The Rochester Institute of Technology in Rochester, New York, has a School of Food, Hotel, and Travel Management. Its chairman, Francis M. Domoy, Ph.D., started a space tourism studies program in 1998. Classes explore how to operate real space resorts from a professional hotel management perspective. Domoy has spoken at space conferences on the subject of operating space resorts.

The University of Houston in Houston, Texas, houses the Conrad N. Hilton College of Hotel and Restaurant Management. Clinton L. Rappole, Ph.D., is developing a real space resort/hotel management program. Being located near the NASA Johnson Space Center, which trains all NASA astronauts, offers his program access to the world's experts on running spaceships and stations.

Both Domoy and Rappole are interested in the concept of space tourism simulations as models for operating real space resorts/hotels. I have had conversations with both, who also see the SimExperience industry as a potential area for employment for their students.

The Space Tourism Society has been conducting research programs on both real and simulated space tourism since 1997. These are directed by experts in the field. We have opened discussions with the administration at Santa Monica College in west Los Angeles, with the goal of starting a simulation academy pilot program in association with their hotel management program. Southern California is an ideal location for a Sim Academy because of its large concentration of aerospace companies, as well as being the center of the entertainment industry, allowing us to draw on a wide range of experts as instructors.

Eventually, some of the Sim Academies could mature into real Space Academies. Who better to train ground crews, space crews, and support staff for real off-world commercial and tourism operations? Of course, this kind of real training will take years, requiring intense academics and on-the-job training, as well as periodic retraining as new equipment and technology become available. Experts who have trained astronauts and cosmonauts and have left government service can be hired to provide training. I have spoken to some who have expressed strong interest in participating in the training of the next generation of commercial astronauts, Simcrews, and staffs.

Sim Science and Research

Another significant advantage and opportunity for establishing and building the SimExperience industry is that we can conduct real social science and design research within the simulated facilities. They will provide living laboratories, allowing us to test design ideas across a large number of people, cultures, and age groups over a long period of time. We can build an extensive database, which will assist us in the design of real off-world facilities and spaceships.

Pioneering scientific research in space psychology, utilizing a totally immersive orbital cruise simulator and the general public as the passengers, has already been accomplished. Harvey Wichman, Ph.D., retired director of the Aerospace Psychology Laboratory at Claremont McKenna College in Claremont, California, directed this pioneering simulation research. It was conducted in 1996 for Bill Gaubatz, Ph.D., retired director/program manager of the *Delta Clipper* program for the McDonnell Douglas Aerospace Space Systems Unit in Huntington Beach, California. Wichman and his students conducted two separate 48-hour orbital cruise simulations, each with eight passengers including student crew members. He acted as head of the mission control team back on Earth, and communicated with the crew and passengers during the simulation runs.

The simulation environment was designed and built by students in his laboratory on the Claremont McKenna College campus. The entirely enclosed spaceship cabin was based on the size of the cargo area of the McDonnell Douglas Aerospace *Delta Clipper* vertical-takeoff-and-landing rocket. Their 400-square-foot simulation area was divided into a sleeping area with bunk beds, along with living, food preparation, exercise, study, and recreation areas. There was a removable table large enough to seat all eight crew members. A portable toilet attached to the simulator offered the same premoistened and prepackaged small towels used by astronauts.

The sim passengers and crew, all wearing light blue astronaut coveralls, were led to the airlock entry by Wichman. Students, college staff, and family members cheered as they entered the sim and the airlock was sealed. Once onboard, the participants went to their bunks to lie down for liftoff. A countdown over the internal loudspeakers was quickly followed by the rumbling sound of the rockets heard throughout the cabin. Once safely in the correct "orbit," everyone left their bunks and assembled in the main cabin. The onboard crew connected with the ground crew and reported on their status. Then, with everyone crowded around a shielded viewport, the launch shield was removed to reveal a beautiful view of Earth from orbit. The footage came from actual video shot by astronauts on Space Shuttle missions. With these and other realistic touches, Wichman and his class achieved a real emotional connection to the space simulation they created, which continued for the next two days.

Hourly activities were scripted by the students and professors to provide opportunities to study interactions between crew and passengers. One mission group received a two-hour, permission briefing that described the purpose of the research and mission. A second group did not receive this briefing and, not surprisingly, experienced more interpersonal conflicts.

The program was a great success, providing valuable data to use in the design of the next-generation of immersive simulation. The passengers uniformly had positive experiences and wanted to be contacted for further studies. They became space experts and celebrities in their own neighborhoods! As a bonus, the students enhanced their abilities to work as a team and interface with professionals in the aerospace industry.

Wichman has continued his pioneering research into space tourism simulations with several other studies. One, sponsored by the Space Tourism Society, utilized a concept I originated where recreation vehicles (RVs) were used as modules for simulated space facilities. A total of 12, or even more, RVs could be connected together to create an immersive environment, simulating a space yacht or cruise ship. The interiors would be designed as spaceships, the windows covered with exterior flat-screen televisions displaying synced views of Earth, space, or the lunar surface. Since the vehicles are mobile, simulations can be conducted around the country.

Other examples of current simulation research include:

NASA and Russian simulation research: NASA has conducted numerous immersive simulations since the early 1960s, some lasting 90 days with crews living in mockups of space station modules, as well as other vehicles and bases. Most of the sims were designed to study the crew's personal interactions and mission performances. Some also tested technologies and techniques for recycling water and waste, with the long-term goal of learning how to construct entirely self-sufficient biospheres in which crews could grow some of their own food. Russian, American, and other space agencies have also conducted numerous scientific simulations.

Biosphere 2 (*www.bio2.com*): Biosphere 1 is our Earth. In 1991, Space Biosphere Ventures opened a $150 million, privately financed, research and testing facility called Biosphere 2. Located on a beautiful site in the Sonoran desert in southern Arizona near Tucson, it is a three-acre glass covered space frame building completely sealed off from the world. It was designed to simulate several of Earth's eco-systems: desert, marsh, ocean, savannah, and rain forest. The profit-making company wanted to pioneer technologies and methods of creating similar biospheres on the Moon and Mars, as well as to do important scientific research into how humans have been affecting the Earth's biosphere through pollution and global warming.

Eight scientists — biospherians — spent two years locked into Biosphere 2 trying to survive within the closed eco-system. However, they were not able to grow enough food and were required to pump in fresh air in order to breathe. They did finish their two-year mission, but there were many critics concerned with the quality of the science. While considered a failure based on its original company goals, it is a beautiful facility well worth visiting. Many business, science, and public relations lessons were learned from the venture which could be helpful to future projects of the same nature.

Mars Society simulated research stations (*www.marssociety.org*): As of mid-2004, the Mars Society had designed, built, and operates two research stations in isolated locations that simulate scientific research on the Red Planet. The Sim missions have six crew members and last between six days and two weeks. According to the Mars Society, "The purpose of conducting such simulated operations is to gain essential knowledge of Mars exploration tactics, human factors issues, and engineering requirements — in short, to start learning how to explore Mars."

The first station was air-dropped in segments on Devon Island in the Canadian Arctic, 900 miles south of the North Pole, a location that most closely resembles the surface of Mars on Earth. Only occupied during the summer, polar bears are the year-round residents and pose a real threat to the scientists who must be armed for protection. The Flashline Mars Arctic Research Station (FMARS) is a $1.3 million fiberglass, two-level cylinder, 30 feet in diameter. The name "Flashline" comes from its primary sponsor. The lower deck has dual airlocks large enough for two people in simulated spacesuits (the airlocks require scientists to wait five minutes before exiting), an Extra-Vehicular Activity (EVA) prep area for six spacesuits, a small lab work area, a hygiene (toilet) enclosure, and a ladder to the upper deck.

It also has a galley and work area with several computers connected to the outside world vi e-mail — the main communications area for the Mars Society mission

The second facility, which opened in December of 2001 in southern Utah, is calle the Mars Desert Research Station (MDRS) and will operate year round. When the call wen out for volunteers to work on building the habitat and to sign up as mission crew members more than 400 people from around the world registered for the selection process.

The National Geographic Society, Discovery Channel, and others have sent film crews out to both Devon Island and the Utah station to do stories. Several magazines and newspapers have also produced stories on these exciting simulated science research facilities

The Mars Society plans on building and operating more of their science station around the world. As of the end of 2003 they have had an impressive track record with more than 150 crew members logging more than 1,800 mission crew days.

I believe the scientific research opportunities will grow along with the commercia potential of the SimExperience industry. I hope that someday there will be a new category of Nobel Prize for breakthrough discoveries in immersive simulations science.

Feel the Future

"The future is where we will spend the rest of our lives." *The TV show* Future Quest, *mid-1990s*

"Feel the Future" is a statement I originated in the late 1990s to describe what it will fee like to participate in immersive off-world themed sims. After all, these sims are precreation of spaceships and off-world facilities, operating in the year 2020 and beyond. It is importan to think of the sims as time machines and not just unique themed locations.

The thematic jumping 20 to 30 years into the future that is a key aspect of sims, provide a rich tapestry of storytelling, character development, and context into which to integrate and demonstrate emerging technologies and software products. The future theme also broaden the market appeal, as well as the pool of potential corporate and university sponsorships.

Fortunately, several of us involved in starting the immersive simulations industry were futurists before we were space advocates, so futurism and forecasting have been a natural part of our planning. I have spoken at the annual conference of the World Future Society (*www.wfs.org*) and can access a wide range of future-oriented think tanks and groups.

These are just a few of the critical path elements that will have significant influence on the future. Many will have to be incorporated into a "Future History" that must be created to establish the context in which the sims operate:

• Orbital development	• Social trends	• Sky tourism
• Communications	• Access to orbit	• Space cruise lines
• Technology	• New materials	• World security
• Exploration	• Sports	• Simulations
• Art and Fashion	• Robots and Telepresence	• Environment
• Artificial intelligence	• Music	• Biospheres
• Genetic engineering	• Food	• Nanotechnology
• Military	• Longevity	• Media

The amount of documented human knowledge doubles about every seven years. Soon it will be doubling every five years. Imagine where these critical elements we have discussed will be in five, 10, and 15 years. All the fields or cultural issues mentioned, and dozens more, need to be researched, discussed, and forecasted into the future in order to create realistic scenarios as the context for the stories and operations of the sims.

There is a subtle yet powerful message incorporated in the off-world tourism vision. It is a positive view of the future. A healthy, vibrant, and successful off-world tourism industry must also mean a successful Earth-based civilization. This positive view of the future is very alluring. It is one of our most potent marketing tools for both the SimExperience industry and the space tourism industry.

The Virtual Experience Community

While you are reading this book, close to a million people around the world are immersed in virtual worlds, role playing a wide variety of characters in a growing diversity of stories. More than 10 million people are participating in these virtual worlds and communities. These sophisticated simulated worlds enable the formation of communities of people linked together by powerful Internet portals and common interest. While the majority of the participants are young, there are people from all age groups, cultures, and economic and educational backgrounds participating.

The role-playing game *EverQuest* is one of the pioneering and most successful games (*www.EverQuest.com*). Now operated through the Sony company, it has more than one million subscribers who pay $20 a year to access the game. There are *EverQuest* tournaments, magazines, and products. Members/participants are now spinning off their own groups of players with their own bulletin boards and special rules. There are many other role-playing games and communities with hundreds of thousands of participants. The computer graphics, artwork, music, and stories are amazing. There is a unique language evolving from these immersive role-playing communities.

I am not an expert in this medium, but I know from talking with experts and players that it is rapidly growing into one of the largest worldwide entertainment businesses and cultural forces. Today the domestic (North American and Canadian) computer game market draws higher revenue than the domestic movie industry. The world's largest video game publisher, with more than $2 billion in annual sales, Electronic Arts, is investing millions of dollars in research, training, and donations to universities, television, cinema, and interactive programs to attract the best new talent to design and program ever more immersive worlds and games. All this creative effort in computer and Internet technologies is making an important contribution to the establishment and growth of the SimExperience industry.

I hope that some of you young readers who are already involved in the virtual world community will take up a new challenge on behalf of the space tourism movement — to pioneer the creation of extraordinary and realistic off-world tourism- and sports-themed worlds and communities. Your efforts could ignite the interest of millions of young people to want to go off world. In 30 to 40 years, they will be of the age when the space tourism industry is blossoming, and they may have the resources to go. I hope that many will have brought their own talents and ideas to our quest.

By the same evolutionary process through which the Disney animation stories an characters came alive in theme parks, virtual world characters could come alive in the nex generation of experience theme parks and themed resorts, and then into totally immersiv Simcenters and Simports. They could stimulate the design and development of actual off world facilities, spaceships, and communities.

In the long run, space tourism- and sports-themed virtual worlds and communitie may become our most effective, pervasive medium for inspiring millions of people t participate in the SimExperience industry.

Conclusion

The quote at the beginning of this chapter, "The tail wagging the dog," underlines m perspective on the overall long-term development of the off-world tourism industry. It wil be the space tourism- and sports-themed virtual and immersive simulation industries tha will provide the catalyst, promoter, and early research funding sources for the off-worl tourism industry. I see wealthy individuals and promotional celebrities who go into Eart orbit through the year 2020, as significant marketing opportunities for the space-theme SimExperience industry.

The space tourism industry will evolve over several decades, while th SimExperience industry could blossom by 2015 into a thriving industry with tens o thousands, then hundreds of thousands of paying Simnauts going "off world" each year.

Many Simnauts will join the growing number of organizations and companies nov pioneering the development of the space tourism movement and industry, contributing thei talents, energy, and funds to accelerate the development process.

Most people interested in space tourism careers will not have aerospace engineerin degrees or the other high-technology backgrounds required to do real rocket science However, they can participate in the simulation industry by coming from a much wide educational and professional background. And, most important, they can get directl involved today.

In the mid-1980s, I purchased a poster showing the evolution of the Mickey Mous character from 1928 to 1986. At the bottom of the poster was a powerful quote — "I onl hope that we never lose sight of one thing … That it all started with a mouse!" (Walt Disney 1954).

PART THREE
THE SPACE TOURISM QUEST

CHAPTER 7
THE SPACE TOURISM MOVEMENT

> "Never doubt that a small group of thoughtful, committed people can change the world. Indeed, it is the only thing that ever has."
>
> *Margaret Mead, anthropologist*

Introduction

Mead's quote is true. A few smart and dedicated people can start a major religion, country, or social movement. By studying history, you realize just how few people it takes to change the world, for either good or evil.

Space tourism in itself is not going to change the world. However, a successful space tourism industry is a powerful tool for creating a positive view of the future and the infrastructure for humanity to expand beyond Earth, creating many new worlds.

Any movement starts with a few people having a common interest or cause. The movement lives or dies based on their ability to communicate, cooperate, raise money, grow their numbers, and attract positive media attention. Through the 1980s, there were only about a dozen of us studying and promoting space tourism. Since the mid-1990s, that number has significantly risen.

The turning point for the movement came in 1996, 10 years after the loss of the Space Shuttle *Challenger* and her crew, when the momentum, positive media coverage, and numbers of space tourism advocates started to grow exponentially. The Space Transportation Association (STA) and NASA cosponsored the first "Space Travel and Tourism Industry Workshop" at NASA headquarters. This year was the first where NASA officially looked at space tourism as a potential industry, I held the first Space Tourism Society planning meetings, the X Prize Foundation made its formal announcement about its $10 million cash prize to stimulate development of private space tourism vehicles, and we also started generating positive articles in prestigious magazines and newspapers. The travel and tourism industry started to take notice of this new potential arena of business.

I believe I am the first person to begin calling our space tourism efforts the "space tourism movement." That was in 1998. I remember wanting to create an identity for our efforts and tested the name during speeches I gave at space conferences and in interviews with reporters. No one challenged the name, so suddenly, after 15 years of effort, we had a space tourism movement going. Clearly the timing was right, and it felt right.

Like any social movement, it is all about people — people who believe in a cause or quest who, over time, convince mainstream society to also believe or act or start something new. The reason I use the term "social" movement is because space tourism is not only about rockets and space stations; it is about creating a new level of powerful personal experiences.

It is about changing the business world's belief that only governments and major aerospace corporations can develop real space ventures. It is about people like you reading this book saying, "I want to get involved. I want to go."

The High Frontier

Since the beginning of the Space Age, some private citizens and groups from around the world have understood just how important the space frontier is to the healthy future of the human race.

The government's space race to the Moon was a great catalyst and inspiration to what became known in the early 1970s as the "Space Movement." The images of astronauts space walking, bouncing on the lunar surface, and driving lunar rovers proved that we can accomplish great things when we make the commitment.

The space movement supported the Space Shuttle program and quested after a Space Station program. Many in the movement and at NASA yearned for a human Mars program to retain large-scale federal funding and to maintain the public's interest and support. However, after the *Apollo* program, the space movement drifted in search of a grand new exciting vision. A vision was required that was more than a federal works program and one with which we could personally connect. A vision that would provide the opportunity for some of us to someday actually go to space.

For many of us, that vision was defined in a book published in 1977 called *The High Frontier: Human Colonies in Space*, by Gerard K. O'Neill, Ph.D., a respected physicist from Princeton, founder of the Space Studies Institute (*www.SSI.org*), and a member of President Reagan's National Commission on Space (1985–86).

In 1974, O'Neill directed a NASA-sponsored summer-long workshop at Stanford University that examined utilizing off-world natural resources such as limitless materials that can be mined from the Moon and asteroids, and constant direct solar energy, to create manufacturing and energy production facilities in Earth orbit. These facilities would reduce the environmental stress and energy needs on Earth by moving some industrial processes outside of Earth's biosphere.

A key part of the plan was to build huge solar power satellites (SPS) in Clarke orbit that would collect direct solar energy 24-hours a day, beaming it to receiving stations on Earth. The energy would then be placed directly into the national grid. A series of these satellites would provide part of the world's energy needs in an environmentally sensitive manner. The SPS concept was created by Dr. Peter Glazer and his associates in the early 1970s. Research into this important concept is still being done in several countries.

O'Neill's plan would require large fleets of reusable launch vehicles for thousands of workers going to and returning from orbit. The workers would require huge space habitats some of which could accommodate 10,000 people with all the animals and farming require to make them completely self-sufficient artificial worlds.

His pioneering work inspired people from around the world to begin thinking of emigrating off Earth to space colonies in the same manner people emigrated from Europe to the colonies in America. One of the most important points made in O'Neill's book centered n the new perception that people from a wide variety of backgrounds could participate in pace development. The book made a direct connection between people and space evelopment. Back in the mid-1970s, this was a revolutionary concept.

Using the *High Frontier* book as a mission statement, Keith and Carolyn Henson unded the L5 Society. L5 is one of five Lagrangian points. These are points within the arth-Moon system where gravitational forces between the bodies are stable, thus allowing ne placement of large colonies in those areas and not requiring large amounts of fuel to be xpended to keep them there. The L5 Society grew into a grassroots movement with private itizens thinking, planning, and acting to develop the first wave of private rocket companies nd businesses designed to commercialize space. The L5 Society had chapters around the orld and held an annual conference.

I joined the L5 Society in 1978 after reading *The High Frontier*. I designed the iteriors of an O'Neill-type space colony as part of my undergraduate architectural studies. was captivated by the concept that I could combine my love for architectural design with iy love for space exploration and become a space architect. That single book changed my fe and those of thousands of other people!

The L5 Society grew and merged, in the early 1980s, with the nonprofit National pace Institute, to become the National Space Society (*www.nss.org*). It is currently based Washington, D.C., has 20,000 members worldwide, and retained the chapter organization f the L5 Society, thus creating numerous local grassroots space advocacy groups. NSS ocuses on promoting the value of space exploration and development to the public. They ave been great proponents of the International Space Station and, since 1996, have ublished articles on space tourism in their magazine *Ad Astra*, which translates as "To The tars."

The Space Frontier Foundation (*www.space-frontier.org*) is more aggressive than SS in promoting the concept of private enterprise in space. They challenge NASA and the erospace industry to be more supportive of private space commercialization. Two SFF ounders, Rick Tumlinson and Bob Werb, arranged for an updated version of *The High rontier* to be published in 1989.

Today you rarely hear much about grand visions of space colonization, even at NSS onferences. That is understandable because the space movement, and now the space urism movement, are focused on near-term goals and fundable projects. However, great oncepts rarely die. O'Neill's concept of a high frontier still simmers in the hearts of many ho were freed by his vision to dream of building strange new worlds.

venting an Industry

s the space tourism movement began to mature from the mid-1990s onward, we began to k ourselves and anyone else who would listen, how do we start the space tourism industry? began by defining that starting point as when NASA, the aerospace industry, and the avel/tourism industry at senior levels, take seriously the potential of private citizens paying illions of dollars of their own money to have a space experience. I knew there were ealthy individuals out there interested in going, and I had a hand in introducing Dennis Tito

to MirCorp at a Space Tourism Society meeting in December of 1999. His flight in 200(
officially started the industry, but even a hundred other citizens like Tito going over man
years cannot sustain an industry and bring it to the mass market. One thousand privat
citizens going each year who generate $1 billion in revenue will justify calling our efforts a
industry.

So, how does one invent an entirely new worldwide industry? First, you study th
history of how similar industries were created. There was once no cruise line industry, n
airline industry, no movie industry, no computer industry. In 1950, there was n
communication satellite industry. In 1980, there was no commercial Internet industry. No
they are all multibillion-dollar industries. By studying how their pioneers create(
developed, and matured their respective industries, we can learn from their advances an
mistakes, and streamline our own efforts.

In most new industries, only a few individuals are credited in history books and b
the general public with being founders. Samuel Cunard was the founder of Cunard, the fir.
cruise line. Henry Ford invented the automobile, creating the automobile industry and th
assembly line. P.T. Barnum is recognized as the creator of the three-ring circus and for bein
America's first great entertainment industry pioneer. Walt Disney is known around the worl
as the creator of Mickey Mouse and for the first theme park, Disneyland. Bill Gates an
Steven Jobs are known as the founders of the personal computer industry.

These visionaries sensed potential and then worked most of their lives to bring it t
reality. Some became incredibly wealthy and powerful; others did not. However, in the
hearts, they all knew they had made a difference, and the world and future generations wer
better off for it.

The Usual Suspects

During the 1980s, more than a dozen professionals began devoting more and more time t
the study and promotion of space tourism as a new industry. I was one of those "dirt
dozen." It would always be the same group speaking on panels at space conferences. As th
media started calling and writing articles and doing television specials about the weir
concept of space tourism, we would all network with each other and usually find ourselv
quoted or interviewed for the same articles and television shows. After a few years, I ha
developed a list for reporters I called "The Usual Suspects."

Below is my Usual Suspects media list. In my opinion, these are the visionaries an
founders of the space tourism movement and industry:

- David Ashford
- Leonard David
- David Gump
- Gene Meyers

- Bob Citron
- Dr. Peter Diamandis
- Jason Klassi
- Tom Rogers

- Dr. Patrick Collins
- Dr. Bill Gaubatz
- Chuck Lauer
- John Spencer

Each of these pioneers has been directly involved in advancing the concept of space touris
as a viable and important off-world industry since the early to mid-1980s, quite often at re

personal and professional expense. Some other exceptional people not on the list because they were not doing media appearances are rocket pioneer Max Hunter, Dr. David Webb, Rick Citron, Makoro Nagatomo, Colette Bevis, Steve Durst, Gregg Maryniak, Dr. Harvey Wichman, Ivan Becky, Robert L. Haltermann, and Sam Coniglio.

The second generation of space tourism pioneers took up the quest in the mid-1990s. Some were involved in private enterprise development in the 1980s or before, but were not yet specifically focused on space tourism. The first international celebrity and real space hero to take up the quest was *Apollo 11* astronaut Dr. Buzz Aldrin. Since 1996, Buzz has made numerous trips and presentations, testified before Congress and at presidential commissions on the economic and social benefits of America's pioneering of the space tourism industry. He has also invested a significant amount of time and engineering talent in creating concepts and designs for orbital access vehicles, some of which could take more than 50 paying passengers into orbit on a single launch. His credibility and space experience is unquestioned and Buzz has brought vital visibility to the movement.

Other astronauts who have made an important contribution to building media and public awareness of the wonders of the space experience through speeches and interviews are Dr. Story Musgrave, who flew six times on the Space Shuttle, and Rick Searfoss, who flew three times on the shuttle, his last flight as commander.

The fact is that today there are many people who are dedicating their careers to the growth of the space tourism movement and industry. I cannot list everyone and, if I tried, I would certainly miss someone because new people join the quest each month. Many of these pioneers are mentioned in the next section, which is a history of the milestones of the movement and industry. The good news is that it is now far too difficult to keep track of all the activities and people around the world engaging in our quest.

As noted elsewhere in this book, Space Adventures is the primary profit-driven company advancing the space tourism industry today. I want to single out Eric Anderson as one of the world's leaders in building the industry, and congratulate him and his talented team for their accomplishments. And, of course, our two heroes who risked their lives, spent millions of dollars of their personal fortunes, and pioneered a path to space — Dennis Tito and Mark Shuttleworth.

Starting the Quest

I think it important to provide you with a historic perspective of how the space tourism movement has evolved since the mid-1960s. Yes, some true visionaries were thinking and writing about private space tourism even before the *Apollo* Moon landings. All of us in the movement today owe them our respect and gratitude.

Therefore, here is my version of the beginnings of the space tourism movement and its key events and accomplishments through the year 2003. It is by no means complete and has a clearly American, West Coast perspective. Peter Wainwright and Dr. Patrick Collins founded the SpaceFuture.com website, which also provides a detailed history of the space tourism movement titled *Our Story So Far*. With their permission, I have adapted and expanded upon information they painfully collected over the years for this section.

Another important source of background information on the history of the space tourism movement and industry was written by Tom Rogers, who is considered the founding father of the space tourism industry. Tom is a highly-respected scientist, engineer, and program manager, with extensive experience at NASA and other government agencies. He was the president of the Space Transportation Association. Tom's paper is titled "Space Tourism — Its Importance, Its History, and a Recent Extraordinary Development." Presented at the International Academy of Astronautics, in May 2000, the paper can be found at the SpaceFuture.com website.

The space tourism movement was lucky to have the Space Tourism Society headquartered in Los Angeles, and the Space Transportation Association's Space Travel and Tourism Division (STTD) headquartered in Washington, D.C. We worked closely together, each having a different emphasis, until the STTD had to close down in 2002 when its key staff retired. STS focuses on research, design, entertainment, and advertising. STTD focused on the political, regulatory, financing, and NASA/aerospace aspects of developing the industry. Between the two groups, we covered most of the important issues and connected with the media to reach the public. We very much hope to revive the operations of STTD or assist in forming another space tourism group based in Washington, D.C.

Space Tourism Timeline

1967   **July:** Baron Hilton, then president of the Hilton Hotel Corporation, presents a paper to the American Astronautical Society titled "Hotels in Space." "… there will be travelers in outer space — and where there are travelers there must be Hiltons," he stated in the paper.

        Kraft Ehricke, a German rocket scientist and one of Wernher von Braun's famous rocket men, also publishes papers on space tourism.

1976   **January:** Rockwell International, the designers and builders of the Space Shuttle fleet, conducts an in-house study looking at building a passenger module that would fill up the entire 15x60-foot cargo bay of the Space Shuttle. Their module design accommodates 74 passengers for short flights to space stations.

1978   Respected space and science fiction author G. Harry Stine writes an important article for *OMNI* titled "Ticket to Space." Harry says, "By the 1990s a trip to an orbiting resort may cost the same as a cruise on the *Queen Elizabeth II*." He advances the idea of private enterprise in orbit and the potential for a huge market for space tourism. In later articles, he skillfully uses comparisons between the railroads and airlines to build an argument for orbital tourism and develops the perception that once you are in Earth orbit you are halfway to anywhere in the solar system.

        Between the late 1970s and early 1980s, some private companies in both the United States and Europe looked at the possibility of commissioning Rockwell to build an additional Space Shuttle, once NASA's four orbital vehicles were completed, and to outfit it with a dedicated passenger module. Designed to spend up

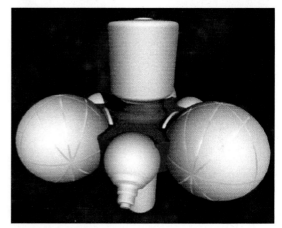

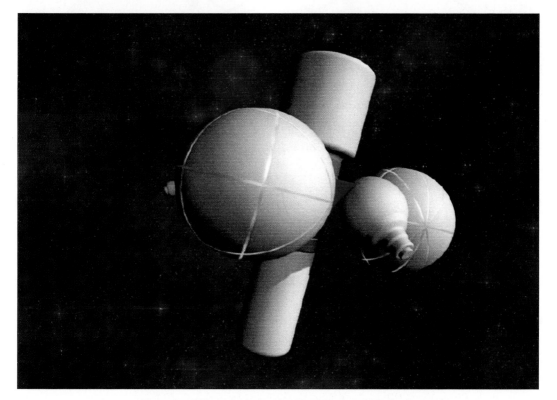

Exterior Views of the Orbital Super Yacht *Destiny*

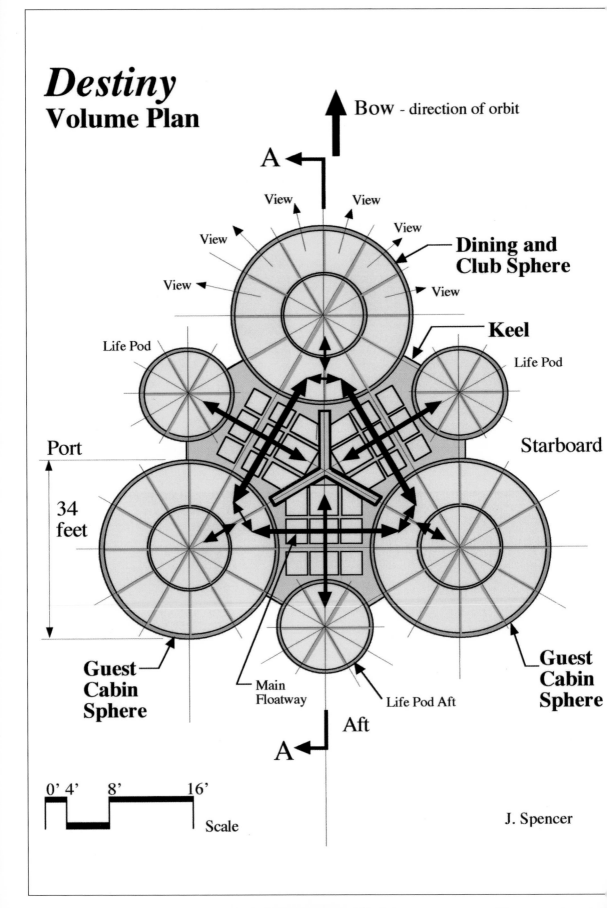

Destiny
Volume Plan

Bow - direction of orbit

A

View · View

View

View · **Dining and Club Sphere**

View

View · View

Keel

Life Pod · Life Pod

Port

34 feet

Starboard

Guest Cabin Sphere

Main Floatway

Life Pod Aft

Guest Cabin Sphere

Aft

A

0' 4'   8'   16'

Scale

J. Spencer

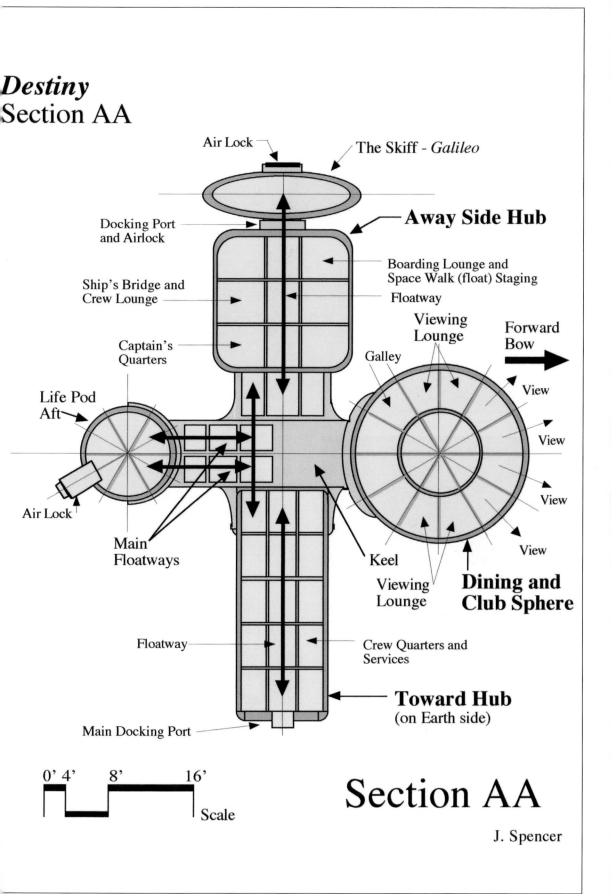

Destiny
Section AA

Air Lock

The Skiff - *Galileo*

Away Side Hub

Docking Port
and Airlock

Boarding Lounge and
Space Walk (float) Staging

Ship's Bridge and
Crew Lounge

Floatway

Viewing
Lounge

Forward
Bow

Captain's
Quarters

Galley

View

Life Pod
Aft

View

Air Lock

View

Main
Floatways

Keel

View

Viewing
Lounge

**Dining and
Club Sphere**

Floatway

Crew Quarters and
Services

Toward Hub
(on Earth side)

Main Docking Port

0' 4'   8'   16'

Scale

Section AA

J. Spencer

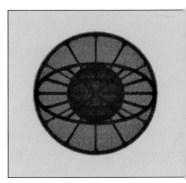

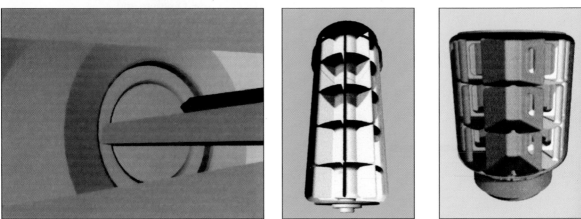

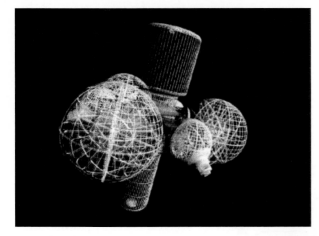

Images show a series of design studies for the interior of the Orbital Super Yacht *Destiny*

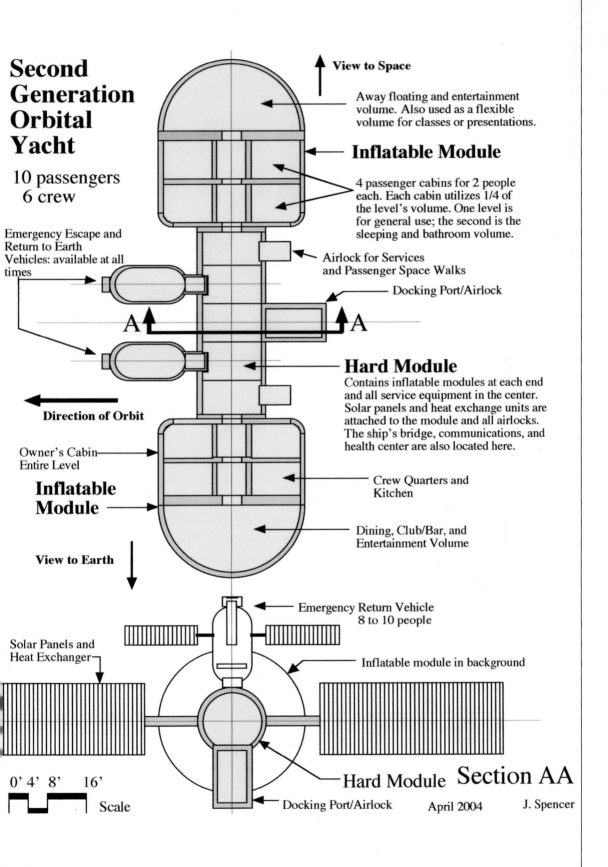

Second Generation Orbital Yacht

10 passengers
6 crew

View to Space

Away floating and entertainment volume. Also used as a flexible volume for classes or presentations.

Inflatable Module

4 passenger cabins for 2 people each. Each cabin utilizes 1/4 of the level's volume. One level is for general use; the second is the sleeping and bathroom volume.

Emergency Escape and Return to Earth Vehicles: available at all times

Airlock for Services and Passenger Space Walks

Docking Port/Airlock

A

A

Direction of Orbit

Hard Module

Contains inflatable modules at each end and all service equipment in the center. Solar panels and heat exchange units are attached to the module and all airlocks. The ship's bridge, communications, and health center are also located here.

Owner's Cabin—Entire Level

Inflatable Module

Crew Quarters and Kitchen

Dining, Club/Bar, and Entertainment Volume

View to Earth

Emergency Return Vehicle
8 to 10 people

Solar Panels and Heat Exchanger

Inflatable module in background

Hard Module   **Section AA**

0' 4' 8'   16'

Scale

Docking Port/Airlock   April 2004   J. Spencer

Examples of the emerging SimExperience industry created by the author. Above is a 10,000 square foot domed attraction called "Mars Base One," designed to be built by a Japanese theme park. Visitors would spend up to an hour touring a realistic second generation Mars base.

The "Mars Resort and Spa," a $20 million venture with a luxury 128-room resort and spa with a 1,000-foot diameter Mars crater to drive a rover on and explore. Most immersive experiences require three parts:

1. An elaborate entry simulation.
2. A place for the guests or explorers to live in.
3. Places to go and come back from.

The Space Resort

Cross the Gantry Access arm into the Space Shuttle

Enter the Space Shuttle and meet the Captain of your flight

Space Resort Master Plan on the Oxnard, California, Ormond Beach site

Countdown and "LIFTOFF"

Watch the docking procedure

An early example of a totally immersive space tourism-themed simulation is the Space Resort, a concept first designed by the author in 1982. Images above show a site rendering including the Space Resort building with a full-scale Space Shuttle stack and gantry tower on the roof. Training facilities and a hotel/convention center are in the background.

Below are images of the wheel-shaped Space Resort on orbit, showing interior and exterior activities. Guests would enter through an elaborate space shuttle and docking simulation, then spend three to five days totally immersed in the orbital cruise experience. Part of the project was built in Japan at Space World that opened in 1991.

Welcome Aboard the Space Resort

Take a Zero Gravity "SPACEWALK"

Fine Dining and a view that is out of this world

Swim in the Athena low-gravity pool

View the stars from your guest bedroom

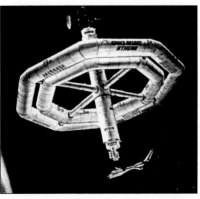

Star tan while you orbit the Earth

The *Athena* Space Resort orbiting 200 miles above Earth

CostaAllegra Crystal Dance Hall

Robonaut shakes hand with astronaut Nanc Currie during a 2003 tes This robot with a human shaped upper torso ha extremely flexible hand and stereoscopic vision. can operate either inside c outside a space statio while the human operate stays safe inside.

The Personal Satellite Assistant will assist astronauts inside the space shuttle or space station. A softball-sized floating information center that will include a video camera, sensors, and a communications platform will respond to voice command. Fans can move the PSA or keep it stationary, as required. It will also monitor interior environment and be on the lookout for leaks.

to one week in orbit, it included sleeping areas, galley, restrooms, and exercise facilities for 24 passengers plus crew. However, as the true cost of Space Shuttle operations began to emerge in the mid-1980s, the concept of purchasing a shuttle dedicated to space tourism became prohibitive.

979 **November:** I form a group called Space Systems Development Group (SSDG) to design a space habitat assembled from 10 Space Shuttle external fuel tanks (ETs). Once in orbit, they would be connected to form a torus (round) configuration, and rotated to produce 20 percent of normal Earth gravity. The ET torus design had been originated by Tom Taylor, a talented civil engineer with extensive experience building projects in harsh and isolated environments such as on the Alaskan pipeline. Of the dozen participants in our study of the ET habitat, *none* were from the aerospace industry. All had the common goal of wanting to go into space and devoted a substantial amount of time to our year-long study. I originated and designed the inflatable interiors of the habitat to accommodate more than 200 people, with some living and working there for up to a year. We planned to have at least 30 people on board as visitors to pay part of the habitat operational cost.

980 **August:** SSDG publishes its self-funded study titled *The Space Shuttle External Tank Habitability Study: 1979–1980,* and begins the laborious process of trying to convince NASA and the aerospace industry to fund further studies. Back in 1980, private citizens just did not propose space projects to NASA or the aerospace community. The approach was revolutionary, and it seemed to frighten some of the aerospace executives and NASA officials with whom we did meet. The common reaction and comments focused around "Why are you here?" However, we did find allies within the agency who supported our ideas and plans.

In the early 1980s, Gene Meyers took up the cause for an ET-derived torus-shaped space facility with Tom Taylor's permission. Gene formed a group called The Space Island Group (*www.spaceislandgroup.com*) and is still actively trying to develop and build the concept. He has made a real contribution to the space tourism movement by exposing many groups and educators to the concept.

982 This is an exciting year for proponents of space tourism. NASA begins the Spaceflight Participation Program to fly private citizens on Space Shuttles. NASA receives more than 11,000 applications. In response to NASA's interest in flying private citizens, Bob Citron, creates an exciting program called the *Space Shuttle Passenger Tour*. Working with Society Expeditions, a well-respected expedition vacation tour operator, their plan is to lease one of the Space Shuttles and take up to 200 paying passengers a year into orbit on a number of dedicated missions in a specialized passenger module. This was the first credible and serious effort to develop real orbital tourism by an experienced expedition tour operator. They hoped to conduct 100 flights between 1995 and 2015.

At the time, NASA is telling Congress and the world that they would be flying shuttles once a month. After exhaustive research and meetings with NASA and Rockwell, Bob realized they could not depend on NASA, so he created a new concept centered on designing and building their own reusable rocket system.

This same year, I create a concept and design called "Space Resort," a $150 million immersive simulation project modeled after the design of Tom Taylor's External Tank torus concept, to provide a three-day simulated luxury space cruise.

Just as I realized that space tourism was the bridge between the federal space programs and private space ventures, I also realized that space simulation ventures were the bridge between financeable space projects in the 1980s, 1990s, 2000s, and real private space ventures in 2010 and beyond.

1983   **January:** Leonard David publishes an article in *Space World* titled "Make Way for Private Citizens in Space." He discusses NASA's interest in flying private citizens on the Space Shuttle by developing a fair selection system. He notes Rockwell's 1970 design of a passenger module for the shuttle cargo bay, along with surveys and market studies that project a large space tourism market potential. Leonard has written several other articles on space tourism for magazines and newspapers, stating elegantly why it is important to develop a space tourism industry and how it can be accomplished.

April: A privately funded study and report is completed by the Space Age Review Foundation, founded by publisher Steve Durst. Titled "The Space Shuttle Passenger Project: A Design Study," the study reviews all the work that had been done to date from around the world concerning private citizen space travel.

1984   The first of a series of papers on single-stage-to-orbit (SSTO), horizontal-takeoff and-landing vehicles for space tourism is published by David Ashford of Bristol Spaceplanes in England.

1985   Bob Citron and Society Expeditions announce a program called Project Space Voyages. They plan to develop a vertical-takeoff-and-landing vehicle along the lines that American rocket pioneer Max Hunter had developed, and, in fact, Max became a member of their team. For several years they collect almost 250 deposits of $5,000 each from around the world toward the full voyage price of $50,000. The deposits are placed in an escrow account. They are unable to raise the development funds, and must return the deposits.

Gary Hudson, founder of Pacific American Launch Systems, publishes and promotes his design for a single-stage-to-orbit vehicle called *Phoenix*.

1986   **January 28:** The Space Shuttle *Challenger* is destroyed during launch, killing its seven crew members and grounding the fleet for nearly three years.

It will be 10 years before private companies in America again begin to pursue flying private citizens in space. However, during this decade, efforts start to examine the economic realities of developing space tourism. Much of this work was conducted by Patrick Collins, Ph.D., in Japan. Patrick is the world's leading expert in the economics of space tourism, having published numerous papers on the subject and other space tourism related subjects. He conducted the first professional surveys striving to establish if there was a real market for space tourism and how much people would pay.

The first book on space tourism is published in Japan by Makoro Nagatomo.

988 **February:** My company, Space Resort Enterprises, receives $1.5 million from Mitsubishi Corporation to form a joint venture called Spaceport Systems Corporation (SSC) to build the Space Resort simulation in Japan and other locations around the world.

989 **January:** The Shimizu Corporation, a major Japanese construction company, presents a paper at the International Astronautical Federation titled "Feasibility of Space Tourism — Cost Study for Space Tour."

At the same time, a well-respected senior scientist/engineer, Tom Rogers, advocates the start of a space tourism industry through articles and meetings in Washington, D.C., with government and regulatory people. He publishes a paper titled "Space Settlements: Sooner Than We Think" in the National Space Society magazine, *Ad Astra*. One of Tom's key points is that "individuals, not governments, may be better suited for the task of creating space settlements." The space tourism movement has been very lucky to have a man of real vision and respectability such as Tom pushing these unique ideas. Many times he remarked how the early pioneers in space tourism were like the brave soldiers of World War I who would throw themselves on the barbed wire so other soldiers could use them as stepping stones to get across to victory. He also would pointed out that this pioneering effort really hurts, and he could show us his scars! In many ways he is our George Washington of the Space Tourism Movement. He even has white hair.

In the late 1980s, Shimizu Corporation forms the Space Systems Division: Space Projects Office to conduct research on orbital hotels.

The Japanese Rocket Society supports the space tourism market research and survey efforts accomplished by Patrick Collins. The JRS also works with the Kawasaki Heavy Industry Aerospace group to develop designs for a 50-passenger vertical-take-off-and-landing, single-stage-to-orbit rocket named the *Kankoh-Maru*. They complete detailed development and operational cost analysis as well as site selection for a new spaceport. Their research continues through today.

990 In the early 1990s, Japanese space tourism pioneers at two of its largest construction companies explore construction methods and potentials in orbit and on the Moon. In the early 1990s, the Obayoshi Corporation develop a number of proposals for large-scale lunar construction projects.

David Ashford, an aerospace engineer in England, working with Dr. Patrick Collins, publishes the second book on space tourism in 1990 called *Your Space Flight Manual: How You Can Be a Space Tourist in Twenty Years*.

December 2: Toyohiro Akiyama, a Japanese journalist working for the Toky
Broadcasting System, is the first private citizen to have his trip to orbit paid for by
private corporation. He blasts off from Russia to the *Mir* space station for eight day
conducting live broadcasts from orbit. The Japanese pay the Russian Space Agenc
$12 million for Toyohiro's ticket.

1991   **May 18:** Helen Sherman, a 27-year-old British chemist, is the second private citize
to have her trip into orbit paid for by private enterprise. After winning a contest ove
13,000 other entrants, she visits the *Mir* space station for eight days. The flight i
called the Juno Anglo-Soviet Mission.

April 12: The Space World theme park opens in Japan, showcasing the world's fir:
full-scale vertical Space Shuttle stack to go on public display, which includes th
shuttle orbiter, external tank, and side-mounted solid rocket boosters. The park is th
world's first space theme park and attracts just under one million visitors a year.

Dr. Bill Gaubatz writes and storyboards a breakthrough report titled "Spac
Is a Place." This position paper can be seen as the Declaration of Independence fc
the space tourism movement. In it, this respected aerospace executive establishes
clear set of priorities for opening the space frontier, with the number-one goal bein
the reduction in cost of access to and return from orbit. Gaubatz quotes the direction
given by President Thomas Jefferson to Lewis of the Lewis and Clark expeditior
"The objective of our mission is to explore … for the purposes of commerce … blaz
the trail and others will follow."

Gaubatz outlines activities and events that could occur in orbit, ranging fror
satellite servicing to assembling a space resort, talk shows, the State of the Unio
Address, and a return to the Moon. Gaubatz blazes a new trail for the then small spac
tourism community. (In 1996, Bill becomes the first senior member of the aerospac
community to join the Space Tourism Society and is named chairman in 2000.)

The U.S. Department of Defense awards a $60 million contract to McDonne.
Douglas Aerospace to develop and test the DC-X reusable vertical-takeoff-and
landing rocket. "X" signifies experimental. The program is directed by Bill Gaubat
and former astronaut Pete Conrad, who was commander of the *Apollo 12* and *Skyla*
2 missions. The private goal of many of the key people on this program is th
development of a space tourism vehicle.

1992   *The Prospects for Space Tourism: Investigation on the Economic and Technologica*
Feasibility of Commercial Passenger Transportation into Low Earth Orbit i
presented at the International Astronomical Federation Congress by Sven Abitzsc
and Fabian Eilingsfeld.

1993   **April 14:** The Japanese Rocket Society starts a study program on the feasibility o
setting up a space tourism business, and establishes its Transportation Researc
Committee to design a passenger launch vehicle. JRS publishes the first special issu
on Space Tourism of its *Journal of Space Technology and Science* (vol. 9, no. 1).

The first market research survey on space tourism is performed by Patrick Collins, Ph.D. A thousand people in Japan answered a written questionnaire, showing that the concept is extremely popular. The results are discussed in "Potential Demand for Passenger Travel to Orbit" and also "Commercial Implications of Market Research on Space Tourism."

August–September: First flights of the reusable VTOL test rocket DC-X take place. The flights demonstrate that operating a rocket-powered vehicle need be no more complex or expensive than operating an aircraft.

994 "The Commercial Space Transportation Study — Final Report" is completed. Conducted by all the major U.S. aerospace corporations, the study for the first time looks into the potential market for space tourism. This was an important step forward within the very conservative aerospace industry.

March: Architect and real estate developer Chuck Lauer presents a paper titled *Space Business Park* at the American Society of Civil Engineers Conference. He had designed a shuttle external tank-derived nonrotating space station that was modeled after Earth-based business parks. One of the potential uses for the facility was to service space tourism. Chuck has done pioneering work in the legal and regulatory areas needed to establish a space tourism industry.

At the Space 94 conference, papers are also presented on the results of the Japanese market research, and on the design of 0-g sports centers by Hazama Corporation, a major construction company.

May: At the 19th International Symposium on Space Technology and Science (ISTS) in Yokohama, four papers from the Japanese Rocket Society study program are presented, including the design of the *Kankoh-Maru*, the proposed JRS passenger launch vehicle that could carry 50 passengers to and from LEO.

October: JRS publishes their second special issue devoted to space tourism in its *Journal of Space Technology and Science* (vol. 10, no. 2). *Considerations on Vehicle Design Criteria for Space Tourism* presents the design of *Kankoh-Maru* at the annual IAF Congress, and a 1/20 scale model of *Kankoh-Maru* is displayed at the Farnborough International Air Show.

995 **March 3:** The year begins with an interesting quote from Newt Gingrich speaking at a National Space Society policy seminar: "I will bet anybody that if we are daring enough, by the year 2015, a major profit center in space will be operating a hotel."

July: I found the nonprofit Space Tourism Society, modeled after the National Geographic Society because they conduct both real science expeditions and research while popularizing it to the general public through the media. STS is the first professional society founded to pioneer the space tourism frontier.

September: Tom Rogers signs an agreement with NASA to hold a joint worksho focused on defining and building a space tourism industry. Ivan Becky, the directo of advanced concepts for NASA headquarters, signs the agreement for NASA an reveals that they had been working on this significant document for a number c years, despite NASA objections.

1996   A growing number of activities worldwide focus on space tourism. It has been 1 years since the tragic loss of the *Challenger* and her crew. Many of us see this yea as the launching of the current wave of space tourism development efforts. This yea saw the beginnings of the mass media taking notice of this strange concept of rea privately funded space tourism.

March: The National Space Society for the first time dedicates an entire issue c *Ad Astra* to space tourism.

A draft paper, "Space Adventure Travel: A Working Paper," is circulated b Gordon Woodcock, one of America's premiere aerospace engineers. He conclude that space adventure travel appears to be commercially promising.

Pioneering scientific research in human factor design, utilizing totall immersive orbital cruise simulation environments and the general public fc passengers, is accomplished by Harvey Wichman, Ph.D., director of the Aerospac Psychology Laboratory at Claremont McKenna College in Claremont, Californi The study is commissioned by Dr. Bill Gaubatz at McDonnell Douglas Aerospac Average citizens respond to newspaper advertisements to participate in 48-hou immersive simulations of a trip to orbit. Two simulations are conducted with si passengers each within a simulator built by the college students. The program is great success with valuable data being recorded for use in the design of a ne generation immersive simulation.

May 15: Results of a Berlin survey are reported in *Prospects of Space Tourism* at th 9th European Aerospace Convention by Sven Abitzsch.

May 18: The X Prize Foundation and Competition are announced at a gala dinner i St. Louis. The site is chosen to commemorate Charles Lindbergh's *Spirit of St. Loui* The X Prize is a $10 million competition to be awarded to the first private compan to finance, build, and fly a three-person vehicle 100 kilometers high — the offici edge of space — and turn it around within two weeks to fly again. The X Prize i designed to stimulate innovative private vehicle development around the world.

Dan Goldin, administrator of NASA, attends the gala and speaks in favor c the program: "I hope my grandson, who is two years old, will be able to go on a tri to a Lunar hotel." Author Tom Clancy also speaks at the dinner. On the spot h donates $100,000 to the foundation. Dr. Peter Diamandis, the founder and chairma of the X Prize Foundation, is also the cofounder of the International Spac University, the Zero Gravity Corporation, and is another founding father of the Spac Tourism Industry.

May: The 20th International Symposium on Space Technology and Science is held in Gifu, Japan. Several papers on space tourism are presented at this prestigious symposium. Barbara Stone, Ph.D., of NASA Headquarters presents "Space Tourism: The Making of a New Industry." Bill Gaubatz, Ph.D., presents "The Road from Delta Clipper–Experimental to Operational SSTO." And Patrick Collins, Ph.D., presents "The Regulatory Reform Agenda for the Era of Passenger Space Transportation."

John Mankins, a senior NASA headquarters staff member, publishes an exciting article titled, "Bed and Breakfast 100 Miles Up!" in the May issue of *Innovation* magazine. In the article he states, "Space tourism will come. It is as inevitable as the Panama Canal and is as irresistible as the development of communication satellites."

June: The American Society of Civil Engineers Space Conference hosts several space tourism papers. In a paper presented by Chuck Lauer, Lauer says, "The 'T' word has come out of the closet."

August 7: NASA headquarters announces to the world that a team of their scientists has discovered the possibility of fossilized Martian life in meteorite ALH84001, a 3.6 billion-year-old rock found in the Alan Hills of Antarctica in 1984. It had been blasted into space when a huge asteroid or meteor hit Mars and, after millions of years traveling in space, fell down our gravity well to Earth approximately 15,000 years ago. While not a space tourism topic, if in the long run this announcement proves correct, this was a historic day, one that could stimulate more interest in space exploration and tourism.

September: The *Kankoh-Maru* model is displayed again at the Farnborough International Air Show, generating several newspaper articles.

September 27: The first NASA/STS cosponsored workshop on Space Tourism is held at NASA Headquarters in Washington, D.C. (See section below.)

November: *Aerospace America* runs an article that Japan plans day trips to space on *Kankoh-Maru* — the first time that a mainstream aerospace journal has done so.

Robert L. Haltermann completes an important paper titled, "Evolution of the Modern Cruise Trade and Its Application to Space Tourism." Bob has a unique background for this work, having been a military officer, NASA officer, and coowner of a travel agency. He becomes the executive director of STA's division of Space Travel and Tourism in 1997.

September 27, 1996

We have arrived! This is an historic day for our space tourism movement. The first workshop on space tourism is held this day at NASA headquarters in Washington, D.C. General Public

Space Travel and Tourism is hosted by Ivan Becky, head of NASA Advanced Concepts, and Tom Rogers, president of the Space Transportation Association (STA). About 50 people from NASA, STA, other branches of government, universities, and private citizens attend, including Gloria Bohan, founder and president of Omega World Travel, and astronaut Dr. Buzz Aldrin. Omega is one of the largest privately owned travel agencies in the world, booking more than half a billion dollars a year. Gloria gives an inspiring speech on the potentials of space tourism from an industry expert's perspective. Among the dozen or so presentations and papers read, I make a speech titled "Wealth from Space Tourism." In that speech, I introduce the concepts of modeling the space tourism industry after the cruise line industry, and announce that I have formed the Space Tourism Society in Los Angeles.

For many of us who have been pioneering the concept of real space tourism, the NASA-sponsored workshop was the first public acknowledgement from the space agency that they were finally listening. Tom, Ivan, and their staffs all deserve special recognition for taking that brave and bold step. This successful workshop set the foundation for a much larger one in early 1997.

1997   **February:** Three papers on space tourism: "Space Tourism: How Soon Will it Happen?" by David Ashford, "Space Tourism: The Surprising New Industry" by Patrick Collins, and "Requirements and Design for Space Tourist Transportation" by Jay Penn and Charles Lindley are presented at the Institute of Electrical and Electronic Engineers Aerospace Conference at Snowmass, Colorado.

February 19–21: The second NASA/STA General Public Space Travel and Tourism Workshop is held at George Washington University in Washington, D.C. Tom and Ivan again chair the workshop, which attracts a more diverse group of 100 participants including many more from the aerospace industry. During the workshop, several working groups address issues such as transportation, finance, marketing, and other key industry drivers. At the end of two days, section chairs present recommendations to the entire workshop. The reports are collected and added to material from the first workshop, and edited into the significant report called: *General Public Space Travel and Tourism: Volume 1 Executive Summary*, Daniel O'Neil, Compiler, Marshall Space Flight Center, Huntsville, Alabama, March 1998 (NP-1998-03-11-MSFC). The final report, *Volume 2: Workshop Proceedings* is published in February 1999 (NASA/CP-1999-209146). (To obtain copies, go to the STA website at *www.spacetransportation.org* or *www.SpaceFuture.com*).

The NASA/STA *Workshop* report notes: "We now see the opportunity of opening up space to the general public — a ('sea change') in our half-century sense that people in space would continue to be very few in number, would be limited to highly trained professionals who, at personal physical risk, would conduct mostly taxpayer-supported scientific and technical activities there under government purview. Now the dream of very many of us during the *Apollo* era that we could someday take a trip to space for our own personal reasons is finally approaching realization. … A survey of the general public showed 42.2% were interested in a space vacation. New lower-cost launch vehicles will one day make the economics of space tourism feasible."

March 15: The inaugural Board of Directors meeting of the Space Tourism Society is held in Los Angeles.

The first International Symposium on Space Tourism is held in Germany in Bremen, with 20 papers presented to 80 attendees, generating very positive national press coverage. The event is organized by Hartmut Muller of Space Tours with DaimlerChrysler Aerospace as the main sponsor.

March 25: The 4th Japanese Rocket Society Symposium dealing with space tourism issues is held.

April 7: *Aviation Week & Space Technology* magazine carries a positive article, "Studies Claim Space Tourism Feasible Based on Papers Presented at the IEEE Aerospace Conference." This is the first time this venerable trade magazine has covered the subject.

April 9: The first national telephone survey in the United States regarding space travel vacations is conducted by Yesawich, Pepperdine, & Brown and the National Travel Monitor. The survey discovers that "40 percent of Americans are interested in an out-of-this-world vacation experience." This represents 80 million Americans!

May 31: The Space Tourism Society holds its first public meeting with 20 people in attendance to review the overall space tourism community situation and to brainstorm ways to build STS. Derek Weber becomes the first STS lifetime member with a $1,000 donation. STS forms an executive committee consisting of Charlie Carr, Jason Klassi, and Tony Materna, with John Spencer as the executive director.

Space Adventures outlines their *Space Flight* and *Steps to Space* programs that also include information on future suborbital flights (available through collateral materials and online at *www.spaceadventures.com*). The true excitement of these space tourism concepts resides in the hearts and the minds of Space Adventures' clients.

July: The *RIBA Journal* publishes a two-page spread on a new space hotel design presented by architects Wimberly Allison Tong & Goo (WATG), the leading resort design company in the world. The work had been done for Gene Meyers based on Tom Taylor's 1979 design concept. Howard Wolff, the vice president of marketing for WATG, presents their design with excellent computer graphics produced by Randy Totel.

Rotary Rocket and Pioneer Rocketplane companies both issue press releases about their advances as private enterprise space access companies. Both propose vehicles designed to take tourists to space.

July 23: The SpaceFuture.Com website debuts.

The JRS Transportation Committee publishes their second report on the development and manufacturing costs of the *Kankoh-Maru* project. It is covered in the *Nihon Keizai* newspaper, receiving nationwide exposure in Japan.

October 5: In his opening speech to the annual congress of the International Astronautical Federation held in Torino, Italy, president Karl Doetsch refers to space tourism as one of the only businesses that will enable the launch industry to grow significantly.

October 6: The first session on Space Travel and Tourism is held at the IAF Congress. A paper from the Aerospace Corporation concludes that passenger carrying rockets are technically feasible today, but that the space industry needs to change its way of thinking to become more in line with the highly successful commercial aviation industry.

November: The Space Frontier Foundation holds its first annual conference in Los Angeles. Tom Rogers, the featured speaker at the main luncheon, briefs the conference on the two NASA/STA workshops concerning the great potential for space tourism to become a large industry. I have the privilege of introducing Tom.

New Type magazine in Japan asks young readers: "What would you most like to do before you die?" The most popular ambition? To travel into space.

This month also sees extensive press coverage of a plan by Zegrahm Expeditions, one of the world's leading expedition cruise operators, to form a new division called Zegrahm Space Voyages. As soon as the vehicles are ready, they plan to offer suborbital flights, followed by orbital flights. A few years later, Zegrahm sells their new division to Space Adventures.

December: The first Space Tourism Society annual meeting is held at the Brentwood Public Library in west Los Angeles. Chaired by yours truly, 50 STS members and associates attend the day-long meeting with Society reports, updates from other space groups and brainstorming for 1998.

1998   **January 1:** Japan's first space travel company, Spacetopia Inc., is founded with the mission of being "dedicated to realizing the new era of popular space travel."

January 25–29: The American Institute of Aeronautics and Astronautics Workshop in Banff, Alberta, on international cooperation in space includes space tourism presentations.

March 25: A press conference is held on Capitol Hill to announce the release of *General Public Space Travel and Tourism*, the final report of a joint study by STA and NASA started in September 1995.

April 15: The Suntory Corporation announces a campaign with Pepsi-Cola Japan in which winners will receive tickets for suborbital flights to space.

April 17–19: At the annual Space Access Society conference in Phoenix, attendees hear updates on the efforts of companies developing reusable launch vehicles and those trying to create a supportive legal environment for space tourism.

April 26–30: Space 98, the biennial space conference of the American Society of Civil Engineers is held in Albuquerque, New Mexico. Featured are sessions on space tourism, space commercialization, space access, and space ports, as well as a special invitation lecture on space tourism by Tom Rogers and Patrick Collins.

May 20: The X Prize Foundation announces the successful raising of $5 million toward its prize target of $10 million and launches the "X Prize Sweepstakes," of which the grand prize is a flight into space. The sweepstakes are sponsored by Space Adventures and First USA Bank, who launch an X Prize credit card featuring space tourism artwork.

In late May, the FAA starts an internal study of extending air traffic management upward to include low Earth orbit.

July 17–26: The Space Tourism Society hosts the first space tourism-themed fair, "Space Fair 98." The 10-day long event is held aboard the *Queen Mary* ocean liner at Long Beach, California, with the theme "Come Celebrate the Dawn of Space Tourism." Staffed by members of STS and the Orange County Space Society (OCSS), the free public fair receives 15,000 visitors and 20 million positive media impressions. More than 30 companies and organizations including NASA, JPL, Boeing, Lockheed Martin, Kodak, GTE (now Verizon), Space Adventures, Universal Studios, Orange County Space Society, HBO, and others provide exhibits. Sponsors include Apple Computer and the *Queen Mary* operating company.

The report "AIAA/CEAS/CASI Workshop on International Cooperation in Space," is published by the AIAA and recommends that "In light of its great potential, public space travel should be viewed as the next large, new area of commercial space activity."

Space Policy Journal publishes the article "Space Tourism: A Response to Continuing Decay in U.S. Civil Space Financial Support," by Tom Rogers.

September: The Japanese Rocket Society forms its Commercial Space Transportation Legislation Research Committee.

September 22: The *Financial Times* newspaper publishes an article on space tourism, "To Boldly Go Away for the Weekend."

September 27–October 2: The IAF Congress is held in Melbourne, Australia. The first meeting of the newly founded International Academy of Astronautics Public Space Travel Group is held. Papers on space tourism dominate the session on economics and commercialization. Papers in other sessions deal with legal issues relating to space tourism.

September 28: *Time* magazine publishes its first article on space tourism, "Vacation in Orbit."

The Japanese Business Federation Keidanren publishes a 10-page pamphlet "Space in Japan," which includes a picture of the *Kankoh-Maru* over a caption stating that they have expectations of space tourism commercializing space activities. The pamphlet ends with the sentence, "Space tourism is expected to give a strong impetus toward the commercialization of space activities."

October 6: The first U.K.-Japan Space Tourism Seminar is held in Yokohama, Japan.

October 9–11: The Space Frontier Foundation Conference in Los Angeles features a panel discussion on space tourism. Notable participants include Tom Rogers, Buzz Aldrin, Eric Anderson, Patrick Collins, Francis Domoy, and myself. Domoy is chairman of the School of Food, Hotel, & Travel management at the Rochester Institute of Technology, which is the first school to begin a series of classes dealing with the issues of operating off-world resorts.

October 20–22: Papers on space tourism are featured at the European Space Agency Workshop on Space Exploration and Resources Exploitation.

NASA Administrator Dan Goldin makes a speech at NASA's 40th anniversary gala dinner: "… in a few decades there will be a thriving tourist industry on the Moon."

October 29: *Mercury* astronaut and former senator John Glenn becomes the first civilian to fly in the Space Shuttle since the loss of the *Challenger*. At 77, he is the oldest person to fly to date. His flight provides a great boost to the image of NASA and the overall American space program, attracting worldwide media attention and reviving discussions about private citizens going into space. However, Glenn follows NASA policy and does not call himself a space tourist, and does not promote flying other private citizens aboard the Space Shuttle.

November 3: Buzz Aldrin speaks for space tourism on *The Late Show with David Letterman*. Buzz does a tour around the country in support of space tourism as we are all encouraged by John Glenn's return to space.

December 12: The second annual STS meeting is held at the Brentwood Public Library in West Los Angeles. Attendance more than triples from the first meeting, to 70 STS members and associates. They gather at the day-long meeting to hear Society reports, watch video tapes, share updates from other space groups, and brainstorm for 1999.

1999   February: NASA finally publishes the second volume of the joint NASA-STS report *General Public Space Travel and Tourism*.

March 1: Rotary Rocket rolls out their *Roton* Atmospheric Test Vehicle at the Mojave Center, to much media fanfare. *Roton* is an entry in the X Prize competition that uses an unusual helicopter-type propulsion system for returning from space to a vertical landing. The company, led by Gary Hudson, goes bankrupt within a year.

March 23: First flight of a reusable vertical-takeoff-and-landing rocket demonstrator is made by engineers from the Institute of Space and Astronautical Science in Japan.

April: A wealthy hotel developer in Las Vegas, Bob Bigelow, enters the space tourism field by forming Bigelow Aerospace, with a publicly stated goal to build a space hotel. Bigelow is willing to spend up to $500 million of his own money to bring the project to fruition.

April 21–23: The 2nd International Symposium on Space Travel is held in Bremen, sponsored by DaimlerChrysler Aerospace. It is again hosted by Hartmut Muller, one of the leading advocates of space tourism in Europe. More than 100 professionals attend, presenting dozens of papers. The symposium receives excellent media coverage in Europe.

May: *Scientific American* magazine publishes a special feature issue titled "The Future of Space Exploration," which includes several articles on space tourism.

May 10: *Space News* carries an article titled "Billionaire Shops for Space Tourism Vehicle," noting that British founder of Virgin Atlantic, Richard Branson, had formed Virgin Galactic to eventually expand Virgin Airlines services into orbit.

June 23–24: More than 100 professionals from around the world attend the first Conference on Space Tourism held in the United States in Washington, D.C. The event is organized by the Space Travel & Tourism Division of the Space Transportation Association.

June 30–July 4: A 1/3 scale model of David Ashford's, Bristol Spaceplanes' *Ascender* suborbital rocketplane is displayed at the Tomorrow's World Live exhibition at Earl's Court Exhibition Center, London.

September 23: The Space Frontier Foundation annual conference in Los Angeles features the first all-morning session on space tourism, chaired by myself.

October 7: The 50th IAF Congress's first full session on space tourism features presentation of eight papers.

November 28: *New York Times* magazine publishes an article by Timothy Ferris, with illustrations by legendary space artist, Bob McCall, titled "Orbiting Resort Hotel."

December 2: The third annual STS meeting is held in west Los Angeles, with 8 members and associates attending, including Dennis Tito. Rick Tumlinson, presiden of the Space Frontier Foundation, makes a special presentation introducing MirCorp a group recently formed to take private control and operate the Russian *Mir* Spac Station, opening it to commercial use, including space tourism. Tito and Tumlinso begin discussing Tito's interest in flying to *Mir*.

2000 **January:** The Millennial New Year breaks with various reports on television, and i newspapers and magazines referring to space tourism. A 90-minute televisio program on space tourism airs on the premiere Japanese news program, *NHK*. It i similar to the news program *60 Minutes* in America.

Norm Augustine, ex-CEO of Lockheed Martin Corporation, predicts tha space tourism will become *the* main space activity in an article in *Aviation Week c Space Technology*.

The formation of MirCorp is announced to the world's news media. Its goa is to commercialize the *Mir* space station.

The American Society of Travel Agents, the premiere travel magazine publishes a major article on space tourism titled: "The Dawn of Space Tourism," writte by Judy Jacobs. This is the first travel publication featuring a story on space tourism

Wired magazine runs a feature article titled "Who Needs NASA," whicl surveys most of the private companies working on vehicles for access to orbit. Th article uses the term "astropreneurs" to describe those pioneering the private use of spa

February 7: *Forbes* runs an article on space tourism, "The Ultimate Trip."

March 2: The papers "The Space Tourism Industry in 2030" and "Orbital Sport Stadium" are presented at the American Society of Civil Engineers Space 200(conference in Albuquerque, New Mexico.

March 17: Astronaut Dr. Story Musgrave and I are featured speakers to an audience o more than 1,000 at the annual conference of the National Tour Association in Pittsburgh NTA is the United States' leading tour operator association. This is the first tim space tourism professionals are paid to speak at a major travel industry conference.

June 1: Three sessions on space tourism are held at the 22nd Internationa Symposium on Space Technology and Science in Morioka, Japan, where presentations are made on Universal Spacelines, the X Prize, airline operations insurance, and certification of the *Kankoh-Maru* for passenger carrying.

June 20: The U.S. Chamber of Commerce announces the formation of their Spac Enterprise Council and its mission to be a broad-based effort to advance the interests of American businesses in the commercial development of space. The new counci is formed in response to many requests for action by Chamber of Commerce members who attended the 1999 Forum on the Future Development of Space.

June 26: The second annual conference on space tourism is hosted by STA-STTD in Washington, D.C., with an all-day discussion of progress toward expanding the space tourism industry. More than 100 professionals from a widening range of industries attend the conference. Discussions about the possibility of Dennis Tito flying to the *Mir* Space Station generate excitement.

July 4: A report on space policy of the U.K. Department of Trade and Industry Select Committee is published. Most important, the committee recommends that the British government review its policy on launch vehicles (quoting from the submission by Space Future Consulting on the importance of space tourism), and that this should not be performed by the British National Space Center (BNSC), but rather by private enterprise. The BNSC had previously blocked funding to Bristol Spaceplanes, Ltd.

July 13: I speak about the space tourism industry at the World Future Society annual conference in Houston.

July–August: The International Space University summer session in Valparaiso, Chile, for the first time includes a design project on space tourism. It was titled "Space Tourism, from Dream to Reality."

September: *GQ* magazine runs a feature article by Steve Kotler on The X Prize titled "Eyes on the X Prize."

October 4: The Space Tourism session at the IAF congress is well attended, with eight papers presented.

October 20: The first meeting on space tourism in France is held by the *Centre National d'Etudes Spatiales* (CNES), the French national space agency.

October 21: An all-afternoon session on space tourism is held at Space Frontier Foundation annual conference in Manhattan Beach, California, chaired by myself and cochaired by Sam Coniglio. The Usual Suspects give updates on promising ventures and new players entering the Space Tourism Movement. Coniglio held the first meeting at the conference for his Space Tourism Club, whose main goal is to attract X Generation interest in space tourism.

October 24: The British government publishes a reply to the Select Committee that does not even mention passenger space travel, but commits to performing an analysis of prospects for a launch market.

October 30: *Time Europe* publishes its first article on space tourism (vol. 156 no. 18) "Destination: Outer Space: As Tourism Booms, Travelers Are Lining Up for a Chance to Sample the Last Frontier."

November: *Popular Science* magazine publishes an article by Dan Cary, "The 10(Mile Club," a general survey of the companies and people pioneering the spac tourism industry.

December: STS holds its fourth annual meeting in west Los Angeles with 8: attendees. Excitement and support is growing for Dennis Tito's impending histori flight to space.

2001 January: In 1968, the Stanley Kubrick movie *2001: A Space Odyssey* was released It has since become *the* classic science fiction movie about space travel. Early in th movie, a private spaceplane run by Pan Am is shown shuttling a scientist to a orbiting space station that includes a Hilton hotel. Following the release of *2001* Pan Am, then the world's most respected airline, launches a First Moon Flights Clul as a publicity stunt. Several thousand people sign up for the first commercial luna flights and proudly carry their cards for years. Unfortunately, the airline does no make it to the real year 2001, but their dream of space tourism has survived Numerous celebrations and a rerelease of the motion picture in major markets occur throughout the year.

Japan Air Line's in-flight magazine, *Winds*, includes a feature article, "The Sky Is Not the Limit," a worldwide survey of all efforts in the pursuit of real space tourism

February 15: I am the featured speaker at the Receptive Services Association (RSA 2001 Tour Summit in New York, where more than 1,000 tour operators hear abou the potentials of space tourism. RSA is the leading tour association for foreign tour: in the United States.

April 12: This day is the 20th anniversary of the first flight of the Space Shuttle, anc also the 40th anniversary of cosmonaut Yuri Gagarin's liftoff in *Vostok 1* as the firs human being to orbit the Earth. To celebrate the historic occasion, "Yuri's Night," worldwide rave/space party is founded by Loretta Hidalgo and George Whitesides and sponsored by the Space Frontier Foundation. Parties are held around the worlc in 40 countries to celebrate the first Yuri's Night. More than 1,500 people attend the Hollywood party. The event is a great success in attracting the interest of the younger generation toward space exploration and tourism.

April 20: I appear on *The Today Show* with retired astronaut Dr. Jerry Linenger in a debate on whether Dennis Tito should fly before the International Space Station is completed. NASA's fight to keep Dennis off ISS generates worldwide media attention.

April 28: The world's first private space traveler, Dennis Tito, rides on a Russian *Soyuz* rocket for an eight-day voyage to the International Space Station. This event officially begins the space tourism industry. His successful flight establishes credibility and interest in private citizens traveling in space, generating more than 1,500 newspaper and magazine articles and more than 500 television news stories.

May 9: Dennis Tito returns to Los Angeles to a press conference hosted by the Mayor of Los Angeles and all the networks. A crowd of 100 cheering friends welcomes him back. The mayor declares Dennis Tito Day in Los Angeles. Dennis appears on the *Larry King Show* on May 15, and is featured on Primetime on May 17. The terms "space tourist" and "space tourism" become part of daily life.

June: The STA-STTD holds its third annual conference in Washington, D.C. More than 140 professionals from around the world attend. Dennis Tito's successful flight is the main topic of the conference. Real progress is being made in Washington with regard to enabling private enterprise to expand the space tourism industry.

July 6: Dennis Tito hosts a cocktail and dinner party on the terrace of his mansion in the hills above Los Angeles to celebrate his flight. His two Russian crewmates, Talgat Musabaev and Yuri Baturin, attend. Almost 200 of the space community join this wonderful evening to formally recognize that the space tourism industry has begun.

October 10: I make a speech at the International Council of Shopping Centers, Tourism, Leisure, and Lifestyle conference in Los Angeles, which is attended by 1,500 top executives and sales managers in the shopping center industry. It is less than one month following the September 11 terrorist attacks on the United States and I state that space tourism represents America's "unstoppable quest to explore new frontiers."

October 19–21: The Space Frontier Foundation's 10th annual conference in Los Angeles features a session on space tourism, chaired by myself and Sam Coniglio. Discussions in the hallways and at meals at the conference focus on space tourism and Tito's successful flight. Dennis is the key speaker at the conference banquet. A 30-year-old Japanese aircraft engineer, Tsuyoshi Saotome, premieres his design for an orbiting space resort called the Crystal Space Palace. His model, display boards, and computer-generated video attract great attention at the conference.

December: The Space Tourism Society moves its annual meeting to April 28 in honor of Dennis Tito's historic flight.

2002   February 21: Dennis Tito is presented the prestigious Americanism Award by the Western Los Angeles County Council of the Boy Scouts of America. He is honored with the award "in recognition of his outstanding career in science, finance, and philanthropy, as well as his historic voyage into space in April 2001." Previous recipients include former U.S. President Gerald R. Ford, comedian Bob Hope, and World War II general James Doolittle.

April 12: The second Yuri's Night celebration has almost 100 cities participating, on all seven continents. Hundreds of people from around the world tune into web casting of the parties.

April 27: The second space tourist, Mark Shuttleworth, a 30-year-old Internet businessman from South Africa, enters the International Space Station for his week-long cruise in Earth orbit. He becomes the first Afro-naut, and is celebrated as the first person from Africa to enter space. Nelson Mandela proclaims him a national hero.

On the same day, the Space Tourism Society holds its fifth annual meeting at Santa Monica College and its first "Space Tourism Pioneer Award Show" in the evening. More than 90 people attend the day-long meeting and deliver a dozen presentations on STS programs and projects, as well as reports from most of the Los Angeles-based space groups. More than 200 people attend the Academy Award styled evening show, featuring Dennis Tito as the guest of honor and main speaker. He receives the first "Dennis Tito Orbit Award." Pioneers of the space tourism movement and industry also receive a total of 16 long-overdue awards. The award show will be held every five years.

May: The popular Japanese magazine, *Brutus*, devotes an entire issue to space tourism, with a focus of lunar tourism. It is the most comprehensive and well produced survey of our new industry to date. An impressive work with excellent graphics and interviews.

June: STS establishes its first international partner in Japan, the Space Tourism Society, Japan. Its founding president is engineer and Crystal Space Palace designer Tsuyoshi Saotome.

Also in Japan, a group of engineers and program managers from NASDA, the Japanese space agency, host a month-long expo dedicated to exploring ideas and designs for space and lunar tourism. Their show generates national media attention.

The Space Travel and Tourism Division of the Space Transportation Association closes down its operations due to retirement of its key officers. The pioneering work done by Tom Rogers and Bob Haltermann stimulated and inspired everyone pioneering the space tourism frontier.

October 11–17: The Zero Gravity Company premieres its plans at the World Space Congress in Houston. The Zero Gravity booth, showing images of private citizens experiencing 0-g inside their aircraft, draws wide attention. They will be the first company in the United States to offer zero gravity aircraft experiences to the paying public.

David Ashford publishes his second book on the subject of space tourism *Spaceflight Revolution*, published by Imperial College Press (ISBN 1-86094-325-X www.icpress.co.uk). The book looks at spaceplanes and other issues relating to building the space tourism industry.

For most of 2002, we witness the first competition for the available third seat on the Russian *Soyuz* spacecraft to the International Space Station. Lori Garver, former executive director of the National Space Society, former Associate Administrator at NASA, and now vice president at DFI International, creates the "AstroMom" program to acquire sponsorship funds to send her into space. She passes all the required physical exams by the Russians. Lance Bass, 23, lead singer of the popular boy band *NSync, attempts to be the youngest person to fly in space. He also passes the required physical exams required by the Russians. Both complete most of the training in Russia at Star City, sometimes training together while vying for the October launch. Both are dedicated but are not able to amass the required sponsorship funding in 2002. They are still questing to go.

003 This year seemed to many to be a quiet year for the movement and industry after the exciting years of 2001 and 2002, marked by the pioneering flights of Tito and Shuttleworth. With economic concerns, terrorism, and the war in Iraq, attention was focused elsewhere.

February 1: Things changed drastically this day as we lost the Space Shuttle *Columbia* and her crew. All Space Shuttle flights are suspended for at least two years. NASA and the Russian space agency struggle to supply ISS using the much smaller and less capable *Soyuz* and *Progress* spacecraft.

　　However, there were significant advances and important new players this year who join the quest or let it be known they have been quietly participating. I will include some space-related events and efforts in this review because they all contributed to keeping public interest in space alive, and in maturing the overall private enterprise space industry.

　　While no private citizen goes off world in 2003, Space Adventures and their Russian partners continue their good work to generate interest from wealthy individuals and corporations who have the resources to fly once the shuttle program is back on track.

February 16: The first public forum on the future direction of human spaceflight, and the implications of the *Columbia* tragedy on private space tourism, is held at the Discovery Science Center in Santa Ana, California. The event is hosted by the Orange County Space Society and is attended by approximately 250 guests. Panel members include Dr. Harvey Wichman of Claremont-McKenna College; Jeff Laube of Northop Grumman; Larry Evans of OCSS; myself; and space educators Richard Shope, Ivor Dawson, and James Busby.

March 1: The Space Frontier Foundation hosts the first-ever summit for many of the leaders of private space groups and companies. The large aerospace firms are specifically not included. The summit is sponsored by Dennis Tito and held at the Riviera Country Club in west Los Angeles. About 50 people attend the one-and-a-half-day event, discussing the effect of the loss of *Columbia* and the opportunity during the rest of the year to contribute to establishing a more progressive direction for NASA and the aerospace industry. Three-time spaceflight veteran and Space Shuttle commander Rick Searfoss, who commanded an earlier *Columbia* mission, gives a moving talk about the space experience and importance of exploration. At the end of the first day, a private reception is held at Dennis Tito's mansion. On the second day, small groups are formed to discuss specific topics and then report back to the overall group.

Ideas from the summit are circulated in Washington, D.C., and other places. At the end of the summit, the first group photo is taken of participants advocating space travel who, individually, have wealth well over a billion dollars.

Dennis Tito, Buzz Aldrin, and several others make trips to Washington D.C. to meet with the U.S. Congress and senior government officials to discuss the regulatory process for certifying private space vehicles and other legal and insurance issues important to expanding the private space industry, including space tourism. For the first time, high-level government officials begin to realize the importance of an American-based private enterprise space industry.

NASA begins the process of reevaluating its strategic mission and goals. The human mission to Mars community meets with Congress to advocate that NASA's main mission should be Mars exploration. The lunar community advocates a human return to the Moon as NASA's primary goal. Some of the push for a lunar return is driven by China's impending success in becoming the third spacefaring nation and its announcement of a human mission to the Moon within 10 years.

March 12: Elon Musk, a highly successful young businessman, who has earned hundreds of millions of dollars, is the featured speaker at the Los Angeles Venture Association's annual Investment Capital Conference. Earlier in the month, he attended the SFF Summit. Elon discusses how he and his associates had succeeded in business and his views on the future of investment. He also discusses his new venture called SpaceX (www.spacex.com), which he is personally financing as a rocket manufacturing and marketing venture. He is focused on small rockets and payloads, but has plans to scale up over the years to human-rated vehicles.

April 26: STS holds its sixth annual membership meeting at Santa Monica College. About 60 members and associates attend the day-long event. Much attention is focused on the *Columbia* accident investigation and issues concerning the International Space Station.

April 28–May 6: I travel to China and South Korea on my second excursion promoting the Space Tourism Society and large-scale space- and Mars-themed simulation ventures. Developers in China are very interested in the space theme.

Other companies continue to make quiet progress on their various space tourism ventures. Bigelow Aerospace makes solid progress on their inflatable tourism module to be added to the International Space Station. The Zero Gravity Company, led by astronaut Byron Lichtenberg and Dr. Peter Diamandis, makes significant progress toward becoming the first private company in the United States to offer the public 0-g aircraft flights. I am engaged to design the interiors of the aircraft. Several of the X Prize contestants make significant progress, including test flying of vehicles and performing major ground rocket tests. Tantalizing bits of information leak out about mega-wealthy individuals investing in space companies.

It becomes known that Jeff Bezos, the founder of Amazon.com, has founded an engineering company in Seattle called Blue Origin to design a reusable vehicle for seven space tourists.

June: *Wired* magazine features an article titled "The Countdown Begins: The Next Space Race Will Be Won by Rich Geeks With the Right Stuff." It is an excellent survey of the dozen new wealthy players who are funding private enterprise space companies. A progress report is made on the X Prize and its leading contender, Burt Rutan of Scaled Composites.

The International Space University (ISU) holds its first session at its new campus in Strasbourg, France. ISU is working to expand its programs worldwide.

June 18: Commercial spaceflight takes a giant leap forward when Space Adventures announces plans to launch the world's first privately funded mission to the International Space Station. Space Adventures secures a contract with the Federal Space Agency of Russia (formerly Rosaviakosmos) to fly two explorers to the ISS aboard a new *Soyuz TM-A* spacecraft. The mission, *Space Adventures-1* (*SA-1*), continues Space Adventures' record of opening the space frontier to explorers other than government astronauts and cosmonauts. The company brokered the flights for the world's first private space explorers, American businessman Dennis Tito in 2001, and the first Afrikaner in space, Mark Shuttleworth, in 2002. In 2004, it was announced that Dr. Gregory Olsen will follow as the third. *SA-1* has the potential to establish several world records.

Space Adventures also continues marketing its suborbital flight program and other Earth-based space-related tours and programs.

August 27: Mars comes closer to Earth than it has been for nearly 60,000 years. There is much media attention for this unique event. It rekindles public interest in space exploration for a while.

October: The first formal meetings in Japan are held for the Space Tourism Society, Japan, chaired by Tsuyoshi Saotome. He and his team build an impressive Board of Governors, Advisors, and Directors. (The inaugural meeting and press conference for STS Japan was held in April 2004).

Popular Science magazine publishes a 130-page special collector's edition titled *Space 2100: To Mars and Beyond in the Century to Come*. It features articles and excellent illustrations on the X Prize and space tourism, including one titled *Heavenly Hilton*. It quotes Rick Tumlinson, cofounder of the Space Frontier Foundation, from a statement he made to Congress where he said, "Space tourism is the killer app … the moneymaker we've all been waiting for."

The Disney Corporation opens its *Mission: Space* pavilion at Epcot Center at Disney World in Florida. Sponsored by Hewlett-Packard, it takes visitors on a wild ride to Mars. Many of the set pieces in the pavilion surrounding the ride are left over props from their movie *Mission to Mars*.

November 25: CBS News airs a special edition primetime show featuring interview with Elon Musk and myself focused on space business development in California and Space Tourism.

December 6: A business enterprise forum is held at Cal Tech, cosponsored by the MIT Alumni Association of Southern California, and hosted by the President of Cal Tech, with Dennis Tito as the featured speaker. He talks about his personal space experience and the progress being made in Washington, D.C., with the Federal Aviation Administration and draft bills in Congress that support private space enterprise and space tourism, as well as his own interest in investing in space companies once these bills are passed.

December 17: The centennial anniversary of powered flight is celebrated at Kitty Hawk, North Carolina. President George W. Bush attends, but disappoints the space community by not taking the opportunity to announce bold plans for NASA and private space enterprise. However, this same day, on the opposite side of the country, history is made by the first supersonic test flight of a privately financed and built aircraft. *SpaceShipOne*, designed and build by Burt Rutan and his company, Scaled Composites, completes a milestone flight over the Mojave Desert in southern California. *SpaceShipOne* and its carrier aircraft, *White Knight*, are clearly the leading contenders to win the X Prize. It is also formally revealed that Microsoft cofounder Paul Allen is providing approximately $25 million in funding for the *SpaceShipOne* flight program.

On a personal note, I finish writing this book, which places a fitting end to what was both a tragic and triumphant year for the space community, the space tourism movement and industry. For many, it felt like a year spent preparing for things to come. I think 2004 and beyond will be far more interesting and exciting.

Dennis Tito's Dream

The space tourism movement has received a priceless gift in the name of Dennis Tito. We could not have gone to a movie studio casting office and chosen a better spokesperson and personality as the first space tourist, or as Dennis rightly prefers to be called, the first private space traveler.

Dennis is a self-made multimillionaire. He grew up in Queens, New York, and was raised by Italian immigrant parents who truly believed in the American dream. As a Boy Scout he formed an attitude that he could accomplish anything if he put his mind to it. He earned a Bachelor of Science degree in Astronautics and Aeronautics from NYU College of Engineering, and a Master's degree from Rensselaer in Engineering Science. At 23, his first job was at the NASA-JPL facilities in Pasadena developing calculations for space probe trajectories. After NASA, he founded Wilshire Associates, a global leader in providing investment management, consulting, and technology services to a wide variety of institutional and high-net-worth investors. The firm manages more than $7 billion for its clients.

A very smart, mature, respected professional in the financial community, he has earned an international reputation for solid judgment in financial matters and donates millions of dollars to medical research and the arts.

For most of his life, he has dreamed of spaceflight and of having the experience of floating in zero gravity while looking at the Earth and stars. Pure and simple, he wanted to go. He was willing to do the training, take the real risks, spend his own money, and fight NASA for the chance to fulfill his dream.

Since returning from orbit, he has been our great spokesperson with the media and the public by promoting the space tourism industry and by conveying his feelings while experiencing space. He has done a fine job and is dedicated to opening space for many more people.

The Space Tourism Society (*www.spacetourismsociety.org*)

The Space Tourism Society's development approach of exploring the beautiful, sensual, and futuristic aspects of space tourism is essential if we are to gain the interest and support of the world public and the finance community. STS is focused on modeling this new industry after the highly successful super yacht culture and cruise line industry, as we have discussed in previous chapters.

I started STS in July of 1995. As of mid-2004, STS and its sister organization, STS Japan, are the only nonprofit professional societies whose sole focus is to develop the space tourism movement and industry. One of our goals is to mentor new STS chapters throughout the world.

STS has about 300 members and associates worldwide. The professional caliber of our membership is outstanding. Our Board of Governors, Advisors, Directors, and Officers are a "who's who" of the space tourism industry. We are honored to have Dr. Buzz Aldrin on our Board of Governors.

Most members join because they have come to their own conclusion that space tourism is the only industry capable of establishing both the technical and financial infrastructure to enable humanity into becoming a true spacefaring species. They find STS through our website, by word of mouth, at conferences, or through the media. There is an exciting diversity of members from a multitude of disciplines, and from more than 10 countries. This diversity is key to our success and to keeping our ideas fresh.

STS's main programs are:

- Developing a "30-year Master Development Plan," whose main goal is to achieve 10,000 tourists going into orbit each year and safely returning to Earth by the year 2030.
- Publishing an STS Journal and expanding our website.
- Holding the "Space Tourism Pioneer Awards Show" every five years.
- Creating an International Design, Finance, and Marketing Symposium.

As with any nonprofit group, we are constantly working on fundraising efforts, including corporate sponsorship and membership programs.

Active STS members feel special among all the other space groups, foundations, and societies because we know we are the vanguard of expanding the interest and support of space exploration and development. We feel a real duty and responsibility to succeed and to do so with a unique grace and style all our own.

Our STS motto is: "Pioneering space tourism experiences for all people."

The STS "Orbit" Awards

One of the most important things STS can do is officially recognize the contributions of space tourism pioneers. This is critical in attracting the interest of new talent, sponsors, and the media. In early 2002, we debuted the "Space Tourism Pioneer Awards Show" and modeled it after the movie industry's Academy Awards. The long-term goal is to have the STS Awards broadcast on television and the Internet.

There are several categories of awards ranging from the highest, the "Dennis Tito Award" for individuals who have paid for their trip and have gone into orbit, to general awards to individuals and groups who have made real contributions to advancing the development of the space tourism movement and industry. The STS membership will eventually vote for the award candidates.

The first "Space Tourism Pioneer Award Show" was held at Santa Monica College' Concert Theater in west Los Angeles on April 27, 2002, one day before the first anniversary of Dennis Tito's launch. Dennis was the featured speaker and first recipient of the award named in his honor. The show attracted more than 200 people and was well received.

The Award Show Committee was chaired by Tony Materna; executive produced by Allison Dollar; and produced, directed, and emceed by Jason Klassi, who also coined the name "Orbit" for the awards.

The two-hour show combined videos with live presentations from the heads of space groups and the presentation of the Orbit Awards. The goal of the first show was to honor the founders and visionaries of the space tourism movement and industry.

Orbit Awards were presented to:

- **Lifetime Achievement Award:** Tom Rogers
- **Space Ambassador Award:** Buzz Aldrin
- **Space Entrepreneur Award:** Walter Kistler
- **International Design Award:** Wimberly Allison Tong & Goo (accepted by David Weisberg, American Institute of Architects)
- **Commercial Achievement Award:** Space Adventures
- **Nonprofit Award:** Space Transportation Association's Division of Space Travel and Tourism (accepted by Robert L. Haltermann, Executive Director)

isionaries Awards were presented to:

- **Research and Conceptualization:** David Ashford, Dr. Patrick Collins, Dr. Bill Gaubatz, and Jason Klassi
- **Public Awareness and Market Packaging:** Dr. Peter Diamandis and David Gump
- **Design and Market Development:** Bob Citron, Charles Lauer, and John Spencer
- **Special Recognition:** Loretta Hidalgo and George Whitesides for creating and producing "Yuri's Night," a celebration of the first human in space

TS will be producing an Orbit Awards show every five years through 2010/15. By that me, we hope to be producing the show annually from then on. We hope many of you ading this book will some day become Orbit Award winners.

onclusion

/e are in the very early stages of space tourism industry development, with decades of dicated work ahead. I find that challenge very exciting and rewarding. I know from talking ith many of the old timers and new supporters of the space tourism quest that we feel a real uty to succeed.

We also know there are millions of people who want to go. Our space tourism ovement is growing. Our job is clear: to give as many people as possible, as soon as ossible, the opportunity to go off world.

Chapter 8
Creating the Market

"Late to bed, early to rise, work like hell,
advertise."
*Wernher von Braun, famed German rocket
scientist and chief designer of the Saturn V*

Introduction

We are not in the *space* business. We are in the *experience* business. Always keep this critical difference in mind.

Through the 1980s and 1990s we persevered and learned how to present the concept of space tourism in a credible manner. After Dennis Tito's voyage in early 2001, the giggle factor finally fell away. However, we must strive continually to refine our vision and widen our media and travel industry connections and influence. We must also inspire the support and direct participation of world-class advertising, public relations, and marketing firms, along with using talented experts from those fields to lead our outreach efforts.

Through the first decade of this new century we must create a vision of the space tourism experience so compelling and beautiful it will capture the interest and support of the media. It is through the media we will reach the people. Our goal is to inspire millions of people to embark on their own quest to go, and to create a verifiable market demand in which the financial community will invest.

By "vision" I mean wonderful designs and images of off-world space tourism experiences and adventures. Space yachts, cruise ships, resorts, and zero gravity sports stadiums shown in beautiful color renderings, virtual reality, and immersive simulations centers, all with happy people from many different cultures and nationalities floating and flying, having the time of their lives, with great views of Earth, space, and lunar landscapes in the background.

Fortunately, space tourism has incredible marketability beginning with the unique nature of the concept and the images. The passion and dedication to fulfill lifelong dreams, and even the tragedies and setbacks, will contribute to the dramatic story we want to tell through the media.

There are several models for building worldwide awareness and market demand for space tourism. The early developers of the cruise lines and airlines faced many of the challenges we now face for space tourism. They found ways to inspire important celebrities to take up their cause and to use their fame to attract media attention. They also used movies and television shows to generate interest from the general public. Worldwide lotteries, with the winners going into space, will also create a direct connection between the public and space tourism.

Making the Space Connection

Most people think of space as some distant and unreachable place. We must change this perception and create a personal connection between people and space.

"Space is only 100 miles away." This statement has proven to be a real attention getter. After hearing me say it, people typically stare for a few seconds until they have processed the information. Then they ask, "Is that true?" Or they say, "Gee, I never thought of it that way."

Most of my meetings and talks are based in Los Angeles, so the next thing I say is "Do you know that you are closer to low Earth orbit (LEO) from where you are sitting than you are to San Diego, 120 miles south of Los Angeles?" This surprises many because most have driven to San Diego and have a feeling of how close it is.

What major city or location is 100 miles from where you live? You have probably driven there many times, so you have a feeling of how close 100 miles is. I call this shift in perspective "making the space connection." Making this connection is critical for the public's ability to accept that someday they can actually go into orbit and that they can become a space tourist.

The next step is to invite people to get directly involved in creating the space tourism movement. I ask questions such as, "How would you cook a gourmet meal in zero gravity?" or "What should orbital fashions look like?" Crowd pleasers always include, "How would you make love in zero gravity or play music or paint?" When they start to answer based on their own areas of expertise or start to ask more questions, then I know that they have just made the space connection in a personal way. For some, this connection will lead to their joining a space group. For others, a new life perspective may be born with the quest to go off world, someday.

A way to make the space connection for many people at once is through television shows and movies centered around space tourism stories. During the late 1970s, the popular television show, *The Love Boat*, featured celebrities taking cruises and finding love on the high seas and in exotic locales. The show revitalized the cruise line industry by exposing the cruise experience to a wider and younger market. We must seek to inspire successful movie and television producers to develop a "Space Love Boat" television series. It could reach millions of people every week with our message.

There are many other ways to make the space connection. The most important issue is that the connection must be made. Without it we have no chance of building a successful space tourism industry. How would you make the space connection for others?

Three Main Markets

As we have explored throughout this book, the space tourism industry will evolve from at least three different markets.

1. Earth-based simulations and tours
2. Off-world yachting
3. Orbital and lunar cruising

By far the largest number of people will participate in the least expensive space tourism experiences of market segment one. Millions of people already participate in Earth-based space-themed experiences. That number will grow as the SimExperience industry matures. There will always be more people having simulated space experiences than real ones.

The next market will be orbital yachting. People are invited to, or rewarded with, a yachting experience. Yachts are social, political, and business tools. Status, pride, ego, and business advancements are the motives for the yacht's existence, not cash profit, although many business deals are consummated aboard yachts and at yacht clubs. The smallest number of people will participate in the exclusive orbital yachting community, but they are the kinds of people who can build the space cruise lines and have influence over planetary affairs.

Following the successful development of the orbital yachting industry will be the orbital cruise line market. When the cost is significantly reduced, safety is significantly increased, and the experience is expanded and refined. Women will be the target market just as they are for the ocean cruise lines. Orbital cruise lines/ports will coexist with the orbital super yachts/clubs, just as their forbearers do on Earth.

A subset of the yachting market/community will be the sports yachts specifically designed for speed and competition in races and regattas. One can forecast huge sponsorships with media attention focused on international space races just as with the ocean-going America's Cup competition.

Specialized market segments will also evolve. The largest could consist of millions of people around the world playing space tourism lotteries trying to win trips off world. The adventure vacation and expedition market will evolve with people going on science- and astronomy-themed trips, hosted by expedition companies or sponsored by nonprofit groups such as the National Geographic Society, the Earthwatch Institute, or the Space Tourism Society.

A large market segment could evolve around 0-g sports in space and 1/6-g sports on the Moon. The merger of adventure travel and major off-world sports events could be the motivation for the largest number of people going off world in the long run.

Forecasting market and social trends 10, 20, even 30 years in the future is a critical part of building the space tourism industry. Fortunately, the art and science of future forecasting has significantly matured since the 1960s. We must enlist the participation of world-renowned futurists such as Alvin Toffler, author of the books *Future Shock* and *The Third Wave*; John Naisbitt, author of *Mega Trends* and *Global Paradox*; and others around the world to guide our long-term planning.

Women — Hear Them Roar

I believe that the space tourism movement and industry will fail unless we engage women as equal partners in its development. The good news is that more women are becoming actively involved. With their guidance we are earnestly reaching out to women's groups and to young women studying a variety of disciplines. With their new ideas and energy we are widening our network of professionals and advocates.

Women make more than 70 percent of the decisions concerning vacation and holiday trips for families and couples. The cruise lines' marketing programs are designe specifically to attract the interests of a woman, and the majority of people in the travel an tourism industry are women.

I purposely study the audiences at travel and tourism conferences where I speal Every time, more than 50 percent of the audience is composed of professional women, man of them senior executives or business owners. These women have a profound influence o the direction of the travel and tourism industry.

Through hundreds of presentations and meetings pitching the space tourisn opportunity, I have noticed that while both men and women respond to the technica adventure, and spiritual aspects of space, separately they slightly favor some aspects other

Men respond to the adventure aspects of space travel, while women respond to th social and personal aspects. Men connect with the adventure, technology, speed, heroics, an team-building to achieve a successful mission. Women connect with the positive an peaceful view of the future we present; the beautiful images of Earth, travel luxury, and sense of heightened individual and spiritual awareness.

Clearly we have two audiences with strong yet similar interests, but with sligh preferences that point to opportunities for marketing space tourism. Excellent! Th observation arms us with the knowledge of how to better present our ideas. We emphasiz space adventure, technology, and team achievement to men, and a positive hopeful view c the future and self-fulfillment to women.

A successful example of a gender-specific presentation was developed by a frien and talented design architect, Lonnie Schorer, and her staff. Lonnie was the senior vic president of the design division for ResidenSea Corporation, which successfully designe and built the world's first luxury resident ship. The *World by ResidenSea* looks like a moder cruise ship, but offers 110 luxury homes instead of staterooms. Passengers are calle residents and own their homes, each costing several million dollars.

Lonnie, who has spoken at some of our space conferences, made a simila observation regarding different perceptions of space travel between men and women whe she presented the ship and home plans to potential buyers. In her case, the men wer interested in the ship and how it was built, how fast it would sail, or how much it weighe while the women were more interested in the home floor plans, the materials, the colors, tl amount of plant life, and kinds of amenities including those for grandchildren.

Lonnie and her staff created two different presentation packages that emphasize different aspects of the same ship. They even used different printing styles for the brochure For women, the brochures were filled with colorful drawings and rendering techniques th softened the technical aspects of the steel ship. In the version for men, more technic drawings with dimensions and facts were used. The gender-specific approach was successf in selling all 110 homes.

Breaking Out!

To branch out of the space, science, aerospace, and science fiction communities that are no talking about space tourism, we must partner with the worldwide tourism industry — fir with the growing adventure and expedition tourism industry, then the yachting community, and then the cruise line industry.

To whom do we market our message of adventure and peaceful spirituality? Everyone, of course. To the young, to middle-aged boomers, and to senior citizens; to the wealthy and those of middle income. Fortunately, by first developing the SimExperience industry, we will bring the experience of off-world tourism closer to the general public both in time and cost.

And then there are kids. Kids love space. It is easy to prove by their television shows, computer games, movies, and the space toys they tell their parents they cannot live without. They are surprised and disappointed that we have not been back to the Moon to build lunar cities and huge space stations. Kids I have spoken with wonder why adults are "so dumb" because they have not been to Mars. This natural connection kids have with space is one of the reasons the various space camps are so successful. They want to go. Right now.

A 10-year-old child in 2000 will be 50 years old in 2040 — just when the real orbital tourism industry will have begun to thrive. He or she will be a perfect candidate for taking a real orbital cruise, and if we have done our awareness-building job correctly, he or she might have taken half a dozen Space Sim Cruises/Adventures and will then, naturally, want to experience the real thing.

Baby boomers were in their teens and 20s when we first landed on the Moon in 1969. The Vietnam War was heating up, there were race riots in America, and popular leaders were being assassinated. The *Apollo* program was one of the bright spots in that troubled time, and boomers feel a special sort of warm relationship to it.

Boomers remember exactly where they were when *Apollo 11* first landed on the Moon, and they like telling their children about the early pioneering days of space exploration. They seem to sense that it is important for their children to know about our great accomplishments and future opportunities in space. However, the boomers are disconnected from the NASA space program of today. The way to reconnect them is through space tourism with the potentials for them or their children to actually go some day.

Seniors tell me they loved listening to *Buck Rogers* and *Flash Gordon* on the radio back in the 1930s and 1940s. They were captivated by the first Moon landings and are happy that these events happened in their lifetimes. They want their grandchildren to learn about our great accomplishments in space, and want them to have the chance to go for themselves. Seniors go out of their way to support our work. Many seniors were inspired when 77-year-old John Glenn and 60-year-old Dennis Tito rocketed into Earth orbit for their experiences off world. In fact, several seniors have told me that one of the brightest spots in their long lives has been the space program. They seem to find real joy and comfort in our expansion off world and all it represents for the future of humankind.

Breaking out to general consumers is only one area of effort. Others include connecting to schools, universities, and their students; to professionals in all fields; and to corporations, governments, and nonprofit groups, demonstrating to them that if they get directly involved in building the space tourism industry, they can benefit themselves and society.

When you feel comfortable talking about space tourism, try talking with your family, then friends, then people at work. I think you will be surprised and reassured by their positive interest. However, be prepared for lots of questions. Space tourism is the kind of subject that naturally generates that kind of excitement. Believe me, there will be a quiz.

The International Perspective

The total population of the United States is only 5 percent of the entire world population. In the future, that percentage will continue to drop. However, as of 2004, I estimate that 80 percent of the people developing the space tourism movement and industry are Americans, although important work is also being done in Japan, England, Australia, and by a few dozen people scattered throughout the rest of the world.

For the same reason it is critical to attract women to the quest of space tourism, we must also strive to attract those in other nations. In fact, space tourism can be one of the shining lights in international cooperation.

The first private citizen to fly on orbit was Japanese. The second was English. The third (who was the first to pay his own way) was an American, and the fourth was a South African. Currently, the Russians are the only country providing access to space for private citizens who can afford to go.

Tourism, by its nature, is the most international of all industries. More than one billion people each year cross national borders. Many people and institutions involved with international issues feel that tourism is a far more valuable and effective tool in easing international tensions than any government propaganda program.

Space tourism and development are the ultimate international border-busters because no country's human-made borders can be seen from orbit, only the natural borders of mountain ranges, rivers, oceans, and desserts. Any major space venture currently require international cooperation and participation. The International Space Station involves 1 countries in its design and construction. More countries hope to eventually join the IS consortium.

By the year 2050, there could be 20 to 25 nations with facilities and ships off worl Someday there could be a United Nations facility in orbit that literally "looks over" th affairs of the planet.

The Sports Connection

I am not into sports. However, when you are searching for industry models and methods create a new market, you look in many directions. To my surprise, the more I studied yac and auto racing, the more connections I saw between the sports industry and space tourist

Sports is a large industry with great influence on the world economy and soci trends. A substantial amount of money is invested in researching new technologie advancing sports medicine and human performance, computer modeling software, spo equipment, shoes, clothing, and so on. In fact, while NASA is always touting its technolo research and spin-off benefits, the sports industry could be doing the same.

We must strive to find allies in the sports industry by interesting sports groups in t long-term potential of sports in space. We must expand their thinking to include how their particular sports could translate into zero gravity or low-gravity sports. They could be the first to host a race in space or a zero gravity NBA basketball game inside the first off-world sports stadium. An extreme sport manned rover race around the entire Moon would draw a huge audience and sponsorship dollars. Instead of the traditional "Baja 1,000," it would become the "Lunar 7,000."

In the near term, new kinds of computer sports games can be created with realistic space physics and environmental considerations to stimulate a child's interest. Space sports-themed television shows and movies would bring sports, entertainment, and space together. With endorsements of sports superstars, we could reach huge audiences.

In the early 1990s, the Hyatt Hotel Corporation paid millions of dollars to air a 30-second television commercial during the Super Bowl that showed a Hyatt Space Resort. It was a fantasy resort design. Hyatt's goal was to show the company's name and logo and say they were thinking about the future.

The main funding source for the sports industry is the media. It seeks hundreds of millions of dollars from corporate sponsors and television commercials. Corporations will pay huge amounts of money to sponsor prestigious sports events like the America's Cup sailboat racing championship and incredible amounts of money to sponsor the Olympics. Someday Olympic games in orbit and on the Moon could generate billions of dollars in sponsorships.

Extreme sports are always pushing the envelope of technology and human performance. Expanding the arena of activity beyond Earth is the next logical step. Credible groups are already researching and developing ways for people to skydive from the edge of space or even from low Earth orbit. Television producers have calculated that more than a billion people would watch a death-defying leap from space.

In the future, say in 50 years or so, a wide variety of sports events will be produced in Earth orbit and on the lunar surface. We could easily discover that sports in space is one of the main market drivers for space tourism. Forming an alliance with the sports industry now could be one of the most productive and profitable steps the space tourism movement and industry can take.

Celebrity Heroes

One of the common tools used effectively by the developers of the early cruise lines, train companies, and airlines was celebrity spokespeople who promoted their young industries. Promoters used royalty, military heroes, politicians, sports figures, and movie stars to build awareness, credibility, and excitement about their new forms of transportation and leisure.

The public is always interested in what the rich and famous are doing, who they are with, what they are wearing, and where they are going. Newsreels and newspapers always showed happy faces of celebrities boarding cruise ships and airplanes to safely carry them long distances to exotic locations. Designed by the industry builders to create interest and confidence in the public and investors, this approach worked and money flowed. Cabins and seats filled, technology advanced, and the world became a smaller place.

The early American space program was extremely fortunate to have the respecte
journalist and television news anchorman Walter Cronkite as a strong supporter ar
unofficial spokesperson. His personal enthusiasm and credibility went a long way
maintaining public interest and political support for the *Apollo* program. When NAS
announced its Spaceflight Participation Program, NASA's first choice was to fly a journalis
Mr. Cronkite was first in line, although later the Journalist in Space program was supersede
by the Teacher in Space project and an educator was chosen by mandate from then-Preside
Reagan.

For many years I have said we need a Jacques-Yves Cousteau for the spac
movement and for space tourism. Someone with the look, the voice, the credibility, and th
passion that Cousteau brought to stimulating interest in the oceans could do the same fo
space.

Dennis Tito has become our respected and eloquent spokesperson. He is our Ne
Armstrong and Jacques Cousteau combined. He is using his new celebrity and spac
experience to promote the future development of the space tourism movement and industr
He has appeared on numerous television programs talking about his space experience. H
mesmerizes live audiences with his unscripted accounts of the freedom and joy of zer
gravity; of the best sleep he has ever had just floating; of being in "heaven," watching th
beautiful Earth silently racing by the viewports; of fulfilling his 40-year dream o
spaceflight; and of coming closer to his family, after facing the real dangers of spaceflight

Dennis is writing a book and a movie about his life's quest to experience spacefligh
He has allowed the Space Tourism Society to create the "Dennis Tito Award" to be presente
to future private space travelers. He is also working with the Russians and Americans o
exciting new commercial space ventures.

Some of our other great spokespeople include Buzz Aldrin. Since 1996, he has bee
a strong advocate of space tourism. He founded the ShareSpace Foundation to focus o
advancing the concept of citizen explorers. He is joined by former astronaut Dr. Stor
Musgrave, who was an active astronaut for 30 years and flew six times aboard the Spac
Shuttle. He retired from NASA in the late 1990s to pursue many creative interests and no
does performance art shows about his space experiences. It is amazing to see the gre
response that Story receives from audiences when he shows images he took himself from th
Space Shuttle. Story tells of his experiences with such passion and reverence that th
audience believes they are right there with him, floating in space. Story gets you as close t
space as does anyone I have ever seen.

One of our most important tasks is to interest celebrities outside of the spac
community to join in our quest and to make it their own, not for a few years but the rest o
their lives. One celebrity hero is actor/director Tom Hanks. He promotes space exploratio
as much as he can and has received the highest possible civilian award for special servic
from NASA. He has made it quite clear that he would love to go off world.

There are dozens of other actors, directors, producers, writers, and others in th
entertainment industry who have participated in space movies and television shows, who ar
supporters of space exploration and development. Most still do not know much about spac
tourism, so it is our job to provide them with the information and create the forums in whic
they can use their celebrity for the benefit of space development and tourism.

A great example of success with celebrities in support of space development was the Space Technology, This Is What's in it for You" public service ads produced by the United States Space Foundation in Colorado Springs (now simply, The Space Foundation). Doug Morrow, an Academy Award-winning producer, spearheaded the project in cooperation with the Ad Council from about 1988–1990, and the ads ran for several years after that.

He teamed unlikely pairs — Willy Nelson with Frank Sinatra, Wilt Chamberlain with Willy Shoemaker, Whoopie Goldberg with Helen Hayes, Tip O'Neill with Barry Goldwater, and Charlton Heston with Gloria Swanson. The copy read: "We don't agree on much, but one thing we do agree on is how space technology is helping society. Space technology, this is what's in it for you."

Those of you in the advertising and media industries could take the above example and strive to create a new version featuring some of today's superstars. Doug Morrow told me he was surprised at how many celebrities wanted to participate in his project out of a love for the dream and a desire to be part of it.

I know from firsthand experience that there are many celebrities interested in space. They are parents and grandparents who want to make their kids proud of them and to be involved in challenging, exciting, and meaningful endeavors. We just need to create the programs and the opportunities for them to participate.

Conclusion

Through the year 2010, the worldwide media is our main audience. We must cultivate the news media, capture their interest, and work with them to promote our maturing vision of off-world tourism and space sports. It is only through the media that we can be heard and seen.

Securing the participation of celebrity spokespeople, and television and movie producers to create space tourism- and sports-themed shows will become the key to our success. We can then build awareness, great expectations, and market demand. This demand can be partially fulfilled by building a successful immersive space simulation industry, which will become our most important media and financing tool.

There is a mystique about space that opens doors in the media. Many journalists and reporters find the subject personally interesting and easy to sell to their producers. Astronauts, and now real space tourists, are still special personalities in society's eyes. This mystique is a critical asset we must exploit to create a positive image and feeling about space tourism. Whenever you have an opportunity, take a reporter to lunch!

CHAPTER 9
THE SPACE TOURISM INDUSTRY
MASTER DEVELOPMENT PLAN

"The best way to predict the future is to invent it."

Alan Kay, Inventor of the overlapping Window interface for personal computers, Popular Science, *Summer 2001*

Introduction

Fortunately, I have a nickname for the long title of this chapter: "The plan."

What is the plan for developing the space tourism industry? I get this question all the time from the media, from excited newcomers to the space tourism movement, and from veterans of the quest. Before you can form a plan, you must have a well-defined set of goals.

Here is our main goal:

- 10,000 tourists going off world annually by the year 2030. That is an average of 27.4 tourists a day, 192.3 a week, and 833.3 a month.

My definition of a space tourist is someone who goes for the experience, not for work, research, exploration, or eventually colonization — someone who pays to go or has a commercial concern pay for them because of winning a contest, lottery, or promotional program. My definition also includes people who are invited to an orbital yacht, yacht club, or who attend a space sporting event.

Defining this primary goal came after many discussions that took place inside and outside the Space Tourism Society. It marks a huge leap from one space tourist per year in 2001 and 2002, to 10,000 per year. But why not? Bold goals and quests stimulate the imagination and passion of those willing to try. One must remember that the goal of sending a man to the Moon and returning him safely to Earth took less than 10 years to accomplish, even though most of the knowledge, technology, and capabilities to achieve that goal did not exist at the time the goal was originally set. We now have more than 40 years of experience going to and from space and the Moon, and far better technologies and tools with which to work. Most important, we will accomplish the main goal through private enterprise and creative innovation, not through government programs.

One person or group is not going to create the plan for the entire space tourism industry. The main point here is that we need a plan. The process of developing it will raise a host of important questions and issues, which are better served in an organized fashion than a chaotic one.

The plan must be created as an international effort. While the majority of space tourism activities currently take place in the United States, Russia, England, and Japan, there growing interest in China and in several other countries. Someday there will be dozens of countries and megacorporations operating in Earth orbit and beyond, so we must strive to define the goals and means to achieve them from a planetary perspective.

The Master Development Plan

There will never be a completed master development plan for the space tourism industry. The plan will always be in the process of becoming. It must be fluid to adapt to opportunities and also to setbacks. I hope there will be several competing master plans because competition is good for stimulating innovation and keeping up the pressure to succeed.

Even after hundreds of years of Earth-based tourism development, there are still new locations for expedition tourism opening up yearly. New forms and concepts of tourism are always being created. The same will be true for off-world tourism. We must remain open minded and sensitive to critical social trends and adjust our plans in accordance to ever changing market trends.

Private enterprise will be the creator, implementer, and operator of these master plans in the same way that they were used to start and build the train lines, cruise lines, and airlines. Some countries' governments will assist their space entrepreneurs, while others will hinder. Do not assume that the United States will be the host country of the space tourism industry. It would be natural to predict that the supposed capital of the free market system would be at the forefront of space tourism. With wealthy people willing to fork over up to $20 million for a week in space, you would think that American entrepreneurs would be lining up to take their money. Instead, Russia is currently the only country actually flying space tourists, and they see the revenue from doing so as a key way to preserve their national space program. China's government has maturing plans for developing a major presence in Earth orbit and on the Moon. Their interest in space tourism is spurred by watching the Russians make millions of dollars flying tourists.

The following is my version of a Master Development Plan. It is based on more than 20 years of pioneering the space tourism frontier and discussions with knowledgeable people from several industries, the general public, and astronauts who have gone off world. Its purpose is to provide an organized discussion tool for the movement, which identifies questions we must face, programs and research we must fund, and alliances we must form, and it requires us to constantly challenge our own assumptions.

At the deepest core of the plan lies the most important point in this book:

We are in the experience business, not the space business.

The uniqueness and quality of the space tourism experience must always be our supreme focus. Every effort, idea, resource, and sacrifice must be made in service of the experience. In the tourism industry, it is the experience that attracts people. If it is a great experience they will come again and again.

Supporting the plan are these three main pillars:

1. Expand the space tourism movement.
2. Define the space tourism market.
3. Finance the space tourism industry.

All three pillars are now happening at the same time. I believe our main focus through the year 2020 must be to establish a vibrant space tourism movement with tens of thousands of talented and dedicated pioneers working closely with the media to promote space tourism. Success will stimulate a market for both entertainment and simulation, as well as actual space tourism experiences. A definable market will enable the finance community to secure the hundreds of billions of dollars needed for the research, technology development, and implementation of the space tourism industry.

Expand the Movement

We are building a movement of highly talented, creative, and motivated people who are dedicating their careers to achieving our main goal and overall quest. Members of the movement must be satisfied with participating in a meaningful quest as their main personal reward. We must create an international community where they enjoy this long-term effort.

A. Adopt a long-term philosophy that participating in creating the movement and industry is the primary personal reward. Those who join the movement have embarked on a lifelong quest in which the process of creating the off-world tourism industry is in itself the reward. The quest is exciting, meaningful, and populated by extraordinary people. Once we accomplish our current goal, we will establish another and then another.

B. Establish a culture that thrives on creativity, diversity, and fun. Creativity is our greatest resource and tool for success. It is critical to enjoy the quest in order to maintain our creative edge and our openness to new and even radical ideas. It is through comradeship and satisfaction from day-to-day accomplishments whereby one lives a good life. Do not let short-term setbacks derail your enthusiasm and dedication. We are smart enough to overcome any challenge given enough time, resources, and creative freedom.

C. Challenge world-class talent and leaders to join our movement. Strive to create a vision of the off-world tourism experience with extraordinary images, designs, music, and poetry. Create a vision that is so beautiful, compelling, sensual, spiritual, and exciting that it will attract the support of the world's brightest and bravest individuals and companies, to embrace the quest for themselves. Challenge them to amaze the world. Reward them with praise and genuine respect. Their love of challenges will push them and the movement forward to success.

D. Become futurists who are always tracking and influencing several key trends. Whil we are a space-oriented movement, we must also be future-oriented. We talk abou achieving our main goal by the year 2030. What will the world be like by then? Ho can we strive to influence key trends over the decades in a dozen tracks that hav powerful influence on our quest? The quote at the beginning of this chapte represents the attitude we should use to invent the future.

Define the Market

We must create a vision of off-world tourism that is so compelling, exciting, and beautifu it will stimulate the interest of millions of people to go. We can accomplish this by buildin the immersive off-world simulation industry, which will provide much of the early fundin needed for key research programs and become our most potent recruitment and marketing too

A. Always work closely with the media to promote our story. Capitalize on the medi attention generated by wealthy private citizens, celebrities, or individuals who ar sponsored to go off world, to build public awareness and interest in real spac tourism. During the pioneering phase, it is the media who are our main marke Through the media we can reach the world public with our story and build marke interest.

B. Attract the interest and support of women. They are our key market. Women mak more than 70 percent of the decisions on where to go for a family or couple vacation. The cruise lines market exclusively to women. We must create a beautifu vision of off-world experiences that will attract the interest and support of women While most men are attracted to the space experience for the adventure, technology and mission accomplishment, most women are attracted to the enlightenin experiences and the future aspects of space tourism. We must inspire more wome to join the quest, become leaders, and with their critical input and talents, design th industry that offers a balance between men's and women's interests.

C. Build a thriving and profitable off-world tourism themed immersive simulatio industry. Millions of people each year experience Earth-based forms of spac tourism by taking zero gravity flights, visiting NASA centers, space camps, spac theme parks and attractions, science centers, space museums, planetariums, an going on astronomy cruises. Tap into this current billion-dollar market as a tool t generate quantifiable market demand studies and funding to build the movement an do critical long-term research and development.

D. Inspire the professional sports industry to create and promote off-world sport activities. The worldwide sports industry generates tens of billions of dollars i annual revenues and reaches huge audiences. They are media-driven and finance via sponsorships and television commercials. Annually they spend millions o dollars on a wide variety of research programs for race cars, racing yachts simulations, materials for equipment, sports medicine, broadcasting means, computer games, and clothing. The industry is always seeking the newest and hottest theme to attract worldwide viewing audiences. Space sports could become the main economic, technical, and market-building driver in the Mature and Mass Market Phases of the space tourism industry.

Finance the Space Tourism Industry

Private enterprise will be the developers and operators of the off-world tourism industry. It must be modeled after the yachting and cruise ship industries, both of which focus on providing high-quality experiences for their passengers. There is a huge amount of diverse research and development that must be financed and conducted in order for the industry to mature. The cost of access to space and off-world operations must come down by orders of magnitude, and the safety must go up by orders of magnitude. These safety needs will necessitate establishment of a Space Guard Service, as discussed previously. This is necessary for insurance companies to provide coverage, which will be required in order to obtain venture financing.

A. Embrace private enterprise as the leading developer and operator of the space tourism industry. NASA cannot participate in commercial ventures and the major U.S. aerospace companies will not risk large sums of money and time on developing this industry. It will only be accomplished by new private enterprise companies and consortiums with an entrepreneurial spirit, like the founders of the cruise lines and airlines. Creative risk takers are the lifeblood of any new industry. They love the challenge and reward of doing something that has never been done before. The aerospace industry will be happy to accept multibillion-dollar service contracts and become the space shipyards on Earth and eventually in orbit.

B. Model the industry after the super yacht and cruise lines industries, not the airline industry. The yachting and cruise line industries focus on providing unique experiences, while the aviation industry focuses on transportation. Work to attract the interest and support of the wealthy and powerful yachting community and the cruise lines to become the owners, developers, and operators of the off-world tourism industry. Use the reemergence of the luxury travel airship industry as an encouragement for their participation and a link between ocean cruising and orbital cruising.

C. Finance long-term research into key technologies and capabilities. As futurists we can deduce what critical technologies, operational capabilities, and guest services will be needed for a thriving off-world tourism industry. Longitudinal research programs are needed so when the financial community finally begins large-scale investments, we are ready to justify their support. Some of the key areas requiring research are:

- Orbital access vehicles and spaceships.
- Robotics, telepresence, and artificial intelligence.
- Self-sustaining biospheres.
- Inflatable structures.
- Advanced long-life batteries.
- Design and testing of personal equipment and clothing.
- Cooking and serving fine meals in a zero or low-gravity environment.
- Health maintenance and emergency care in zero gravity.
- Regulations, treaties, and insurance.

D. Establish the Space Guard Service and private space academies as soon as possible. Model the Space Guard after the U.S. Coast Guard to provide the same services in orbit and beyond. This is a critical service required for private enterprise to insure its ventures and, therefore, acquire financing. We must also establish private space academies for training of personnel in the operation of off-world facilities and spaceships, as well as Earth-based infrastructure. These academies can be modeled after maritime academies with the addition of courses in space science, off-world operations, and cruise ship-style hospitality.

There are many other things we must accomplish on an international level in order to achieve our main goal. However, the most important thing we must do is to maintain our own enthusiasm, while keeping our minds open to novel ways for accomplishing our goal in a joyful and creative way. The following sections explore ways to implement our Master Development Plan.

The Art of Finance

One of the first questions businesspeople ask me about space tourism is, "How much will it cost?" Businesspeople have the perception that any space project with people going off world will cost tens of billions of dollars and can be paid for only by nations. That perception is false. There are several designs for vehicles and businesses that would not cost billions of dollars to send people to and from orbit. Therefore, one of our most important objectives is to change that perception and to inspire the most talented and brilliant financial minds to accept the challenge of financing the space tourism industry.

Any reasonable venture can be financed if there is enough profit to be made, enough power and prestige to be gained, and if the risk is spread over many players and the debt service is spread over at least 50 years.

In the mid-1980s, a $10 billion mega-merger caused major international headlines. Today, only $40 to $50 billion deals cause media interest. According to *Forbes* magazine there are more than 500 individual billionaires throughout the world. Bill Gates, the cofounder of Microsoft, is worth more than $40 billion, which is a truly astounding level of wealth. There are millions of millionaires around the world.

The good news is that every year more wealthy individuals are getting directly volved in financing parts of the growing private space enterprise industry, including space urism ventures.

As of the end of 2003, I know of three multibillionaires who are directly involved in ace tourism-oriented ventures by investing millions of their own dollars: Paul Allen, ofounder of Microsoft; Jeff Bezos, founder of Amazon.com; and Richard Branson, founder f Virgin Atlantic (although Branson has not been as active over the past few years). There e more than a dozen multimillionaires who are also now directly involved in spending their oney on private space ventures.

How does large-scale venture financing work? Large-scale expensive projects are aid for by a variety of methods including taxes, bonds, corporate and government uarantees, private equity, debt, sponsorship, and a combination of any of the above. The ey part of any financing formula is the length of time it will take to pay back the loans. ypical development projects require at least 55 years on their loans. The longer the loans, ne less the annual debt service costs, but the more interest is paid.

There are several huge construction projects in the works around the world, each osting more than $40 billion. However, the cost is not really an issue in financing these rojects. The revenue and profit generated over the life of each project is what really counts. inancing a $100 billion venture, such as an orbital yacht club, several orbital yachts, and ozens of orbital access vehicles, and their support infrastructure, is well within the range of ne right kind of international financing and development consortium.

While most people would not think of bankers as "creative," high-level investment ankers are very creative and extremely smart. I work with investment bankers and have ersonal experience in finance and fundraising. I have been in finance meetings that are just as xciting and creative as design meetings. Over the years, I have developed a philosophy that ou can design your financing in the same way you design your architectural or media project.

I believe we are a decade away from the time when we can realistically begin the Wall Street" type of financing process for real space tourism ventures. Significant research eeds to be done in a wide variety of areas. A new generation of reusable orbital access ehicles must be developed, tested, certified, and insured. A clearly definable market must e created and demonstrated to satisfy the financial community. We have a lot of work to do, ut we also have growing interest in the financial world. They are watching our progress ith genuine interest.

olitical and Regulatory Issues

his book does not attempt to address the important issues of politics and regulations overning space development and operations. It does not address the legal or insurance ssues associated with launching, recovering, or operating spaceports, orbital access ehicles, orbital facilities, and spaceships. Fortunately, there are dedicated individuals and roups pioneering these issues, which must be addressed if we are to be successful. Several ew books addressing the policy and regulation requirements of space development are eing produced by this publisher and are available through Apogee's website. Many timely apers can also be found at the spacefuture.com website.

Great Expectations

We can manifest great expectations of wonderful off-world experiences by connecting those special people who have already gone with those who are interested in going. Through great expectations we create a desire to go that translates into a quantifiable market demand.

As designers, developers, and operators of space tourism experiences, we will be required by the market *we* have created to fulfill these expectations and, therefore, be empowered to bring our original visions into reality. This is a classic example of a self-fulfilling prophecy. It is also a constant reminder that we must protect our focus on the quality and uniqueness of the off-world experience.

However, we must be careful not to promise too much too soon. We must build realistic expectations and be clear that it will take decades and huge sums of money to establish access to the experiences described in this book. There have already been a number of disappointments caused by eager new participants in the development process who made exaggerated claims and then failed to deliver. The investors and the public demand honesty and if you are honest with them, you will acquire their support.

Space Age Wealth

We are fortunate that the emerging new wealth in industrialized nations is based on technology, information, and creative content. Those creating and managing this new wealth grew up in the Space Age. They are accustomed to space missions and take for granted that we can access Earth orbit most of the time.

This Space Age wealth population is between 30 and 60 years old. They grew up watching the *Apollo* and Space Shuttle missions, the assembly of the International Space Station, successful robotic exploration missions to every corner of our solar system, and the *Star Trek* television series and space movies. From firsthand experience, I know space tourism intrigues this group. The idea that they can go someday blows their minds and creates real excitement. Many times, I have seen their eyes light up when they realize that an entirely new and fantastic life experience is waiting for them. Many also comprehend the awesome business, prestige-building, and promotional potential, and they want in. As members of this Space Age wealth culture become even more powerful and effective in the economic and political arenas, they will become the foundation on which we will succeed.

Ego and Prestige

The strong egos of the early pioneering astronauts and cosmonauts are legendary in the space mythos. As fighter and test pilots, they had to have strong egos in order to survive. Today, the challenge is different. Many space travelers have a Ph.D. or extensive engineering, scientific, and medical experience. But strong egos and determination are still required to successfully compete in the selection process and to excel during the years of training. Finally, anyone who goes must have the willpower to master the fear of launching into orbit and the violence of reentry. The *Challenger* and *Columbia* disasters proved that you can lose your life questing for the space experience. Pioneering space tourists had to have strong egos to create the wealth that enabled them to pay the millions of dollars, do the cosmonaut training, and courageously risk their lives.

Getting the space tourism movement and industry to where we are today has required strong egos of the founding fathers and mothers — egos strong enough to withstand the laughter in the early days, knowing in the long run we were right and would succeed.

Today, prestige is slowly replacing strong egos as the key motivator. The prestige a group obtains from success can bring competing egos together into a team to achieve a collective win. Having studied the super yacht industry and yacht racing culture for several years, it is crystal clear to me that we have a truly winning strategy focusing on positioning orbital tourism as a prestigious endeavor.

The prestige of owning the world's first orbital yacht, being a founding member in the first orbital yacht club, winning the first orbital yacht race, and being the first private citizens to yacht around the Moon will be an irresistible challenge to some of the ultra wealthy and powerful.

When the space tourism movement has succeeded in establishing the goal of owning an orbital super yacht as one of the most prestigious achievements in history, we will automatically attract those powerful people and companies who can build the space tourism industry.

Leading Leaders

Our strategy for attracting the best of the best to our quest requires the current leaders of the space tourism movement to learn how to lead leaders. How? Give them the vision, direction, and extreme challenges, then reward them with prestige and honor. They will then lead themselves.

I saw a television interview with a very successful movie producer who said his philosophy is to challenge talented people to "amaze him." I thought that was a brilliant approach to dealing with talented and powerful people working for you in a way that gets them to produce remarkable results.

A good example of leading leaders is Dr. Peter Diamandis' effort with his Ansari X Prize Foundation. He and his associates set the goal, established the prize, and challenged the rocket scientists and space entrepreneurs around the world to win that prize. With more than 25 teams working, Peter has marshaled a strong talent pool of leaders leading themselves.

The Space Tourism Society's Space Tourism Pioneer Awards are designed to recognize people who have made real contributions to the space tourism movement and industry. I cannot emphasize enough how important it is to challenge people to a noble quest, one that calls upon them to bring forth their very best to achieve success, one they know will serve a higher purpose, and then recognize them for their efforts.

The Creative Spirit

I believe our movement's most powerful resource is our individual and collective creativity. You never know when an individual's or group's brainstorming will originate a new concept

or idea that will change or enhance our Master Development Plan for the better. Tho
creative leaps are welcome. We need outside perspectives that challenge all establishe
parameters, while helping to create exciting new ones tailored to the uniqueness of spac
Exploring the creative frontier of space tourism design and development is open to everyon

We must take this creative frontier spirit and apply it to all aspects of our work. W
are competing for attention, talent, and all forms of resources, so we must be highly origin
in order to stand out in our promotional, marketing, design, recruitment, and fundraisir
efforts.

We must create channels whereby new ideas easily reach the leaders of ou
movement and industry. A student from India may have a breakthrough idea that we mu
see. The president of a major aerospace company might have an insight that we must hea
A retired astronaut or military officer might invent a new technology we must support.

We must promote easy exchange of ideas as a cornerstone of our movement an
industry. By doing so, we significantly enhance our potential for success, and we foster
culture and environment that stimulate creativity and innovation.

Conclusion

Even the best strategies of today will change and mature in five years, 10 years, 20 year
The world will change; attitudes will change; technology and capabilities will significantl
improve. A country or megacorporation could step forward and proclaim that building th
space tourism industry is a priority program. Our goal of 10,000 space tourists a year b
2030 could be easily surpassed. Alternately, technical, economic, environmental, or terrori
challenges and setbacks could push the date of achieving this goal off by a decade or mor

Everything we need to achieve our main goal is within our capability to acquire
to create during the next three decades. We have the vision, the mission, the dedication, an
the talent to succeed. We are also empowered with the knowledge that our quest is importar
for the long-term prosperity of humankind, and we as individuals and as a movement ar
making a difference.

While this book and Master Development Plan propose several concept
approaches, and timelines, there are limitless options. There are few right or wrong answe
at this early stage of development, in fact, the questions are more important than th
answers.

After reading this book you will probably have your own opinions and ideas. I woul
love to hear about them. Please e-mail me at *JSSDesign@aol.com*. Attend a Space Touris
Society meeting and those of other space and future groups. Express yourself!

Creating and building the off-world tourism industry is a long-term process. Once w
have achieved our current main goal, we will move on to our next main goal — 10,000 luna
tourists a year! Then 10,000 Mars tourists a year! And I can imagine that, someday in th
distant future, our descendants will be brainstorming with their artificial intelligenc
partners on ways to take tourists to the stars.

CONCLUSION

> "Far better it is to dare mighty things, to win glorious triumphs, even though checkered by failure, than to take rank with those poor spirits who neither enjoy or suffer much, because they live in the gray twilight that knows not victory or defeat."
> *Theodore Roosevelt, April 10, 1899*

No one laughs at the idea of a space tourist anymore. That reaction stopped with Dennis Tito's flight and has been replaced with a more sober, determined, practical, and professional attitude. The space tourism industry has been born and is steadily growing into a mature effort. It is growing because people are willing to feed it money, time, and vision — and to risk their lives to have unique life-changing experiences.

Eventually, the wealthy and powerful will build orbital yacht clubs and orbital super yachts, then host prestigious space yacht racing to satisfy their ego and pride. The cruise lines and airship lines will extend their services into orbit and to the Moon. The sports industry will establish and promote off-world sports and sports stars, stimulating technological advancement and widening the market. Movies and television shows will be shot off world, and human exploration of Mars will capture the world's attention for a brief time. Mars tourism will capture the interest of wealthy adventurers.

I am not a writer, so writing this book was the hardest thing I have ever done. It took five years and getting up very early hundreds of mornings so I could have a few hours of quiet, focused time to work. But I felt it was a duty to trace the birth and growth of this effort, to create a Master Development Plan, and that doing so could make a difference in peoples' lives. I truly hope it does.

I know I have made a small difference in founding and building the Space Tourism Society, writing this book, making hundreds of presentations and speeches on space tourism, and designing dozens of off-world simulation and entertainment projects. It is a good feeling.

However, our quest is not an easy one. It will continue to be a real challenge for the next few decades as the world focuses on economic, environmental, and security problems. But overcoming challenges is what makes success so sweet and rewarding.

I would like to share a personal experience that has been a source of strength and comfort to me. One night in the mid-1990s, I was channel surfing when I came upon the image of a knight in armor battling his way through a horde of creatures guarding a door. Victorious, the knight steps through the door, his sword raised and prepared for battle, but instead finds himself in a quiet library lined with rows of bookshelves, and a single table and chair by a burning fireplace. Shocked, the heroic knight searches the library for someone to fight. Finding no one, his battle face begins to show signs of confusion and worry. Standing by the table, he notices an engraving on the wall that reads, "If ye are truly dedicated to the

cause ye will read this library and emerge a better champion." Clearly the knight prefers th
excitement and challenge of battle. He looks longingly at the door, wanting to step back ou
but bows his head and leans his sword against the table. Taking off his helmet an
breastplate, he walks to the first bookcase, pulls the first book from the top row witho
reading the title, sits down, and begins to read. The television screen fades to black with
few words in white: "There are many kinds of heroes."

I turned off the television and just sat there for a long time. I never knew who create
the commercial or why, but I knew their message had enhanced my life. I came to call th
experience "The Knight in the Library" and whenever I am frustrated over our slow progre
in building the space tourism industry, I think of this knight and his dedication to his caus
As those powerful words stated, "There are many kinds of heroes." You can be a hero b
bringing your own voice, ideas, and dedication to the space tourism quest.

I am more encouraged and inspired today than ever before. You can be a part of th
grand quest — a truly noble quest filled with excitement and opportunities. The process o
creating a successful off-world tourism industry will be the key economic and technologic
driver enabling the human species to evolve into a real Solar System Species.

Some decades in the future, I hope to be floating in line, listening to a man or
woman in front of me complaining to the robotic attendant about how long it is taking
process their luggage. I will look out a large viewport at the beautiful Earth and smil
knowing that all the years of dedication and hard work have paid off. We have created a
irritated space tourist!

Do you want to go?

Epilogue

"A man does not become old until regrets replace his dreams"
Unknown

I was watching the evening news in early November 2002. The commercials had started so I had muted the television. Still watching, however, I half saw a few scenes of a beautiful sunset by a lake, a man listening to tapes (Russian language lessons, I found out later), and the man getting in shape by running and working with a trainer. We see him drinking energy fluids and watching his big-screen television, which happens to have a scene of a Lunar Rover bouncing across the Moon's surface. That caught my full attention.

He then looks at his computer and is checking the weather in Moscow, packs his luggage, then heads to his office to sign a contract with corporate types discussing the document all around him. A 747 flies over and we land in a foreign airport with a small woman holding a sign looking for "Mr. Douglas." We see a quick tour of Red Square in Moscow with many dour-looking Russian officials.

Next we are on a bus heading into the woods. The first hint of something special comes here as we see a passenger riding the bus with "Mr. Douglas" who looks suspiciously like a Russian cosmonaut. We pass a guard gate where the man in the gatehouse barely glances up as we pass through the checkpoint.

Then my attention is riveted as the first true indication of his destination appears. His escorts walk him through the zero gravity water training facility at Star City. By then I was getting excited. Could this be a space tourism television commercial?

A close up on his face as a visor drops and locks into place, a smile of anticipation is seen, along with a twinkle in his eye. Next is a view in the flame trench of a *Soyuz* booster undergoing ignition, followed by the rocket arcing up into the cold Russian sky.

In space, after transferring to a space station, "Mr. Douglas" ecstatically floats free. A large porthole is behind him. The first words come on the screen: "When your kids ask where the money went…" We float toward the porthole to see the black of space and the brilliant blues of the Earth below. His digital camcorder in hand, "Mr. Douglas" floats toward the window, his face illuminated with pure joy and awe. Then the final words: "Show them the tape." By now I had goosebumps. Fade out to the name, "Sony."

I could not believe it. A primetime television commercial modeled after Dennis Tito's flight. A minute later, the phone started ringing and kept ringing for hours. I saw the premiere of this commercial by sheer luck. What was so exciting was that I, and all my space tourism associates, had not heard any advance buzz about this commercial. Sony thought the theme of space tourism would be a good one to launch a new high-end product line, and so they used it.

It was strangely gratifying to have had nothing to do with the commercial. It felt like the concept of space tourism was finally alive and growing on its own. It was like watching your child taking its first independent step. Very cool.

Resources

Website Addresses

Ansari X Prize Foundation . . . . *www.xprize.org*
Apogee Books . . . . *www.cgpublishing.com*
Buckminster Fuller Institute . . . . *www.bfi.org*
Challenger Learning Center . . . . *www.challenger.org*
Bigelow Aerospace . . . . *www.bigelowaerospace.com*
Blue Falcon Editing . . . . *www.bluefalconediting.com*
Futron Corporation . . . . *www.futron.com*
Global Friendship Through Space Education . . . . *www.gftse.org*
Institute for Accelerating Change . . . . *www.accelerating.org*
International Space University . . . . *www.isunet.edu*
Joseph Campbell Foundation. . . . . *www.jcf.org*
LunaCorp, Inc. . . . . *www.lunacorp.com*
Mark Shuttleworth . . . . *www.africaninspace.com* and *www.tsf.org.za*
Mars Society Simulated Research Stations . . . . *www.marssociety.org*
National Space Society. . . . . *www.nss.org*
Orange County Space Society . . . . *www.ocspace.org*
Red Planet Ventures, Inc.. . . . . *www.redplanetventures.com*
Rick Searfoss, Colonel, USAF Retired. *www.astronautspeaker.com*
Society Expeditions. . . . . *www.societyexpeditions.com*
Space Adventures, Inc. . . . . *www.spaceadventures.com*
Spaceflight Revolution, by David Ashford . . . . *www.icpress.co.uk*
Space Frontier Foundation . . . . *www.space-frontier.org*
SpaceFuture.com . . . . *www.spacefuture.com*
Space Island Group, Inc. . . . . *www.spaceislandgroup.com*
Space Tourism Society . . . . *www.spacetourismsociety.org*
Space Tourism Society, Japan. . . . . *www.uchutaiken.com*
U.S. Space Camp . . . . *www.spacecamp.com*
World Future Society . . . . *www.wfs.org*
Yuri's Night . . . . *www.yurisnight.net*
Zero Gravity Corporation. . . . . *www.zerogcorp.com*

E-Mail Contacts

John Spencer . . . . *JSSDesign@aol.com*
Karen Rugg. . . . . *Krug4space@aol.com*
Larry Evans. . . . . *Mach25@sbcglobal.net*
Andrea Howe . . . . *Andrea@bluefalconediting.com*

ACRONYMS

AS	American Astronautical Society
I	Artificial Intelligence
IA	American Institute of Architects
IS	Artificial Intelligence Sentinel
IAA	American Institute of Aeronautics and Astronautics
SCE	American Society of Civil Engineers
STA	American Society of Travel Agents
STC	Association of Science-Technology Centers
NSC	British National Space Center
BS	Columbia Broadcasting System
EAS	Confederation of European Aerospace Societies
NES	Centre National d'Etudes Spatiales (French national space agency)
EO	Chief Executive Officer
EV	Crew Exploration Vehicle
NN	Cable News Network
C-X	Delta Clipper-Experimental
SV	Deep Submersible Vehicle
T	External Tank (Space Shuttle system)
SA	European Space Agency
VA	Extra-Vehicular Activity
AA	Federal Aviation Administration
MARS	Flashline Mars Arctic Research Station
BO	Home Box Office (cable movie network)
A	Interactive Architecture
AF	International Astronautical Federation
EE	Institute of Electrical and Electronic Engineers
SAS	Institute of Space and Astronautical Science
SR	Institute for Scientific Research
SS	International Space Station
STS	International Symposium on Space Technology and Science
SU	International Space University
PL	Jet Propulsion Laboratory
RS	Japanese Rocket Society
SC	Johnson Space Center (Houston, Texas)
SC	Kennedy Space Center (Cape Canaveral, Florida)
EO	Low Earth Orbit
M	Lunar Module (*Apollo* program)
MDRS	Mars Desert Research Station
MEB	Mars Expedition Base
M	Maneuvering Module

MSF	Mobile Ship Function
MSFC	Marshall Space Flight Center
MSNBC	Microsoft/National Broadcasting Company (cable news service)
NASA	National Aeronautics and Space Administration
NASDA	National Space Development Agency (Japan)
NBA	National Basketball Association
NOAA	National Oceanic and Atmospheric Administration
NSI	National Space Institute
NSS	National Space Society
NTA	National Tour Association
OAV	Orbital Access Vehicle
OT	Orbital Taxi
OTH	Orbital Transfer Hub
OTS	Orbital Transfer Ship
OZ	Orbital Zone
OZP	Orbital Zoning Plan
PST	Pacific Standard Time
RIBA	Royal Institute of British Architects
ROI	Return on Investment
ROV	Remotely Operated Vehicle
RSA	Receptive Services Association
RV	Recreational Vehicle
SAM	Space Awareness Movement
SAS	Space Access Society
SFF	Space Frontier Foundation
SGS	Space Guard Service
SIG	Space Island Group
SPS	Solar Power Satellite
SRB	Solid Rocket Boosters (Space Shuttle system)
SSC	Spaceport Systems Corporation
SSDG	Space Systems Development Group
SSI	Space Studies Institute
SSTO	Single-Stage-To-Orbit
STA	Space Transportation Association
STS	Space Tourism Society
STS	Space Transportation System (Space Shuttle)
STTD	Space Travel and Tourism Division (part of STA)
USA	United Space Alliance
USAF	United States Air Force
USL	Universal Space Lines
USSC	U.S. Space Camp
VTOL	Vertical Take-Off and Landing
WATG	Wimberly Allison Tong & Goo

INDEX

ABOUT THE AUTHOR

John Spencer is a space architect who has built a career that is a balance between the design and finance professions. He is a pioneer in what he calls "The Design Frontier." He creates, designs, and develops his own space tourism-, Mars-, and future-themed immersive simulation and attraction projects, while serving as a conceptual designer for some of the world's largest corporations.

As a real estate developer and site master planner, he coined the term "Experience Park" in the early 1980s. His 1982 Space Resort design and development project (to take a three-day simulated cruise into low Earth orbit) matured into the Space World theme park built in Japan, opening in April 1991. He is the creator of the first *Star Trek* theme park design and of the Science Fiction Hall of Fame. John coined the terms "SimExperience" and "Simnauts" to describe a new generation of totally immersive recreation simulation experiences he is currently designing, including orbital cruising and Mars exploration. He is the founder and president of Red Planet Ventures, Inc.

Since 1978, he has been recognized as a pioneer in the field of outer space architecture. He completed the first interior designs of the Spacehab module in 1983/4, which have now flown on orbit more than a dozen times. In 1995 he was awarded the Space Act Award and the Certificate of Recognition from NASA for innovative architectural design on the International Space Station, now in Earth orbit. In 2000, he was awarded the Space Humanitarian Award by the United Societies in Space and *Apollo 11* astronaut Dr. Buzz Aldrin. In 2002, he was awarded the Space Tourism Pioneer ("Orbit") Award from the Space Tourism Society.

In 1997, under contract to TRW, he visited Christmas Island in the South Pacific and created the first launch complex conceptual master plan for the Japanese Space Program. In 2000, John completed the first interior design for the Universal Space Lines *Space Clipper*. In 2003, he designed the interiors of the Zero Gravity Company aircraft.

John designed the interiors for the *Aquarius* Underwater Laboratory, built in 1984 by NOAA, which is still in operation. For the National Science Foundation, he and his team provided all the design work for an Antarctic Science Base assembled in 1985 and 1986, and originated the concept for a Mobile Field Laboratory, built in 1990.

He has been a proponent of space tourism using the cruise ship industry model since the early 1980s, and is the founder and president of the nonprofit Space Tourism Society.

Since 1997, he has been designing the world's first orbital super yacht called *Destiny*, premiered in this book.

John has been quoted in *The Wall Street Journal, Scientific American, Space News, Popular Science, Men's Journal, Los Angeles Times*, and more than two dozen other magazines and newspaper articles on Space Tourism. He has appeared on *The Today Show*, CNN, ABC, CBS News, and on more than a dozen other television programs including the Discovery Channel, PBS, Japanese NHK, and the Learning Channel. He has a professional (1980) and masters degrees (1982) in architecture from the Southern California Institute of Architecture.

What people have been saying about Apogee's award winning space series.

"A unique reference providing details available heretofore only to researchers with access to the archives."
Library Journal

". . . (the series) will serve as an invaluable reference tool for aficionados of human spaceflight." Astronomy

"The package is guaranteed to put space enthusiasts into orbit." Today's Librarian

". . . budding Tsiolkovskys, Goddards, and von Brauns will devour each title." Booklist

"Highly recommended for space buffs who want detailed information on these flights." Choice

". . . a space freak's dream." National Post - Canada

"This series is highly recommended . . . A bargain at twice the price!" AD ASTRA

". . . must represent some of the best value for money possible for any space enthusiast." Spaceflight - British Interplanetary Society

"The most ambitious look at our space program to date" Playboy

". . . a mine of useful information and well worth getting." Astronomy Now

"Enthusiasts will enjoy having this historical record." Publishers Weekly

Quotes from Space experts

"Your work is important to the future of space exploration." Chris Kraft - Apollo Flight Director

"I believe your books are the best possible resource you could use." Guenter Wendt - Pad Leader

"The book is riveting. It is of immense historical importance." Sir Patrick Moore

"The CD's are marvelous, I'm afraid I'm going to lose a lot of sleep!" Sir Arthur C. Clarke

"...of incalculable value to anyone interested in the future of the human race." Spider Robinson

"You've done it again: Found the coolest stuff and put it into one, big, bounteous package unlike anything else that's available. And the DVD is super!" Andy Chaikin

"Robert Godwin has produced yet another gem in his now classic series of historical books on NASA's space missions." Dr Pascal Lee

"We cannot overrate this book and DVD." Boggs Space Books